T0214219

Lecture Notes in Computer Science 12554

Founding Editors

Gerhard Goos
  Karlsruhe Institute of Technology, Karlsruhe, Germany
Juris Hartmanis
  Cornell University, Ithaca, NY, USA

Editorial Board Members

Elisa Bertino
  Purdue University, West Lafayette, IN, USA
Wen Gao
  Peking University, Beijing, China
Bernhard Steffen
  TU Dortmund University, Dortmund, Germany
Gerhard Woeginger
  RWTH Aachen, Aachen, Germany
Moti Yung
  Columbia University, New York, NY, USA

More information about this subseries at http://www.springer.com/series/7412

Xiahai Zhuang · Lei Li (Eds.)

Myocardial Pathology Segmentation Combining Multi-Sequence Cardiac Magnetic Resonance Images

First Challenge, MyoPS 2020
Held in Conjunction with MICCAI 2020
Lima, Peru, October 4, 2020
Proceedings

Springer

Editors
Xiahai Zhuang 🆔
Fudan University
Shanghai, China

Lei Li 🆔
Shanghai Jiao Tong University
Shanghai, China

ISSN 0302-9743 ISSN 1611-3349 (electronic)
Lecture Notes in Computer Science
ISBN 978-3-030-65650-8 ISBN 978-3-030-65651-5 (eBook)
https://doi.org/10.1007/978-3-030-65651-5

LNCS Sublibrary: SL6 – Image Processing, Computer Vision, Pattern Recognition, and Graphics

© Springer Nature Switzerland AG 2020, corrected publication 2021
This work is subject to copyright. All rights are reserved by the Publisher, whether the whole or part of the material is concerned, specifically the rights of translation, reprinting, reuse of illustrations, recitation, broadcasting, reproduction on microfilms or in any other physical way, and transmission or information storage and retrieval, electronic adaptation, computer software, or by similar or dissimilar methodology now known or hereafter developed.
The use of general descriptive names, registered names, trademarks, service marks, etc. in this publication does not imply, even in the absence of a specific statement, that such names are exempt from the relevant protective laws and regulations and therefore free for general use.
The publisher, the authors and the editors are safe to assume that the advice and information in this book are believed to be true and accurate at the date of publication. Neither the publisher nor the authors or the editors give a warranty, expressed or implied, with respect to the material contained herein or for any errors or omissions that may have been made. The publisher remains neutral with regard to jurisdictional claims in published maps and institutional affiliations.

This Springer imprint is published by the registered company Springer Nature Switzerland AG
The registered company address is: Gewerbestrasse 11, 6330 Cham, Switzerland

Preface

Assessment of myocardial viability is essential in the diagnosis and treatment management for patients suffering from myocardial infarction (MI). Different cardiac magnetic resonance (CMR) sequences can image and provide unique information of the heart. These sequences include the late gadolinium enhancement (LGE) CMR, which visualizes MI, the T2-weighted CMR, which images the acute injury and ischemic regions, and the balanced steady-state free precession (bSSFP) cine sequence which captures cardiac motions and presents clear boundaries. Combining these multi-sequence CMR data can provide rich and reliable information with regards to the pathological as well as the morphological information of the myocardium.

MyoPS 2020 provides the three-sequence CMR, i.e., bSSFP CMR, T2 CMR, and LGE CMR, from 45 patients. All the clinical data has received institutional ethic approval and has been anonymized. The data released here has been pre-processed using the multivariate mixture model method, to align the three-sequence CMR images into a common space and to resample them into the same spatial resolution. The training images are provided with gold standard labels, including left ventricular (LV) blood pool, right ventricular blood pool, LV normal myocardium, LV myocardial edema, and LV myocardial scars. MyoPS 2020 also intended to present an open and fair platform for various research groups to test and validate their methods on these datasets acquired from the clinical environment. The aim is not only to benchmark various myocardial pathology segmentation algorithms, but also to cover the topic of general cardiac image segmentation, registration, and modeling, and raise discussions for further technical development and clinical deployment.

A total of 16 papers were accepted and presented at MyoPS 2020, and are published by Springer in this LNCS volume. MyoPS 2020 was held in conjunction with the MICCAI 2020 international conference. MyoPS 2020 was scheduled to be held in Lima, Peru on October 4, 2020, but finally was held through a virtual conference management platform due to the COVID-19 pandemic. The readers can find more information about MyoPS 2020 at the website:
http://www.sdspeople.fudan.edu.cn/zhuangxiahai/0/myops20/.

We would like to thank all organizers, reviewers, authors, and sponsors for their time, efforts, contributions, and support in making MyoPS 2020 a successful event.

October 2020

Xiahai Zhuang
Lei Li

Organization

Chairs and Organizers

Xiahai Zhuang	Fudan University, China
Lei Li	Shanghai Jiao Tong University, China
Fuping Wu	Fudan University, China
Xinzhe Luo	Fudan University, China
Yuncheng Zhou	Fudan University, China
Jiahang Xu	Fudan University, China

CMT - Springer Conference Submission/Publication System

Lei Li   Shanghai Jiao Tong University, China

Webmasters

Xiahai Zhuang   Fudan University, China
Fuping Wu   Fudan University, China

Challenge Website

http://www.sdspeople.fudan.edu.cn/zhuangxiahai/0/myops20/
https://zmiclab.github.io/projects/myops20/

Sponsors

We would also like to thank our sponsors:
https://www.deepvessel.net/
https://sds.fudan.edu.cn/

Contents

Stacked BCDU-Net with Semantic CMR Synthesis: Application to Myocardial Pathology Segmentation Challenge

Carlos Martín-Isla[1]([⊠]), Maryam Asadi-Aghbolaghi[2], Polyxeni Gkontra[3], Victor M. Campello[1], Sergio Escalera[1,3], and Karim Lekadir[1]

[1] Departament de Matemàtiques & Informàtica, Universitat de Barcelona, Barcelona, Spain
carlos.martinisla@ub.edu
[2] Institute for Research in Fundamental Sciences (IPM), Tehran, Iran
[3] Computer Vision Center, Univeritat Autónoma de Barcelona, Barcelona, Spain

Abstract. Accurate segmentation of pathological tissue, such as scar tissue and edema, from cardiac magnetic resonance images (CMR) is fundamental to the assessment of the severity of myocardial infarction and myocardial viability. There are many accurate solutions for automatic segmentation of cardiac structures from CMR. On the contrary, a solution has not as yet been found for the automatic segmentation of myocardial pathological regions due to their challenging nature. As part of the Myocardial Pathology Segmentation combining multi-sequence CMR (MyoPS) challenge, we propose a fully automatic pipeline for segmenting pathological tissue using registered multi-sequence CMR images sequences (LGE, bSSFP and T2). The proposed approach involves a two-staged process. First, in order to reduce task complexity, a two-stacked BCDU-net is proposed to a) detect a small ROI based on accurate myocardium segmentation and b) perform inside-ROI multi-modal pathological region segmentation. Second, in order to regularize the proposed stacked architecture and deal with the under-represented data problem, we propose a synthetic data augmentation pipeline that generates anatomically meaningful samples. The outputs of the proposed stacked BCDU-NET with semantic CMR synthesis are post-processed based on anatomical constrains to refine output segmentation masks. Results from 25 different patients demonstrate that the proposed model improves 1-stage equivalent architectures and benefits from the addition of synthetic anatomically meaningful samples. A final ensemble of 15 trained models show a challenge Dice test score of 0.665 ± 0.143 and 0.698 ± 0.128 for scar and scar + edema, respectively.

Keywords: Cardiac magnetic resonance · Myocardial pathology segmentation · Deep learning · BCDU-Net · LGE · bSSFP · T2

C. Martín-Isla and M. Asadi-Aghbolaghi—contributed equally to this work.

© Springer Nature Switzerland AG 2020
X. Zhuang and L. Li (Eds.): MyoPS 2020, LNCS 12554, pp. 1–16, 2020.
https://doi.org/10.1007/978-3-030-65651-5_1

1   Introduction

Myocardial viability assessment is key in the diagnosis of patients suffering from myocardial infarction and ischemic heart disease, among others. Cardiovascular magnetic resonance (CMR) is a well-established imaging technique that provides anatomical and functional information of the heart. Multiple sequences with different properties can be acquired, registered and combined to obtain a complete viability assessment. Late gadolinium enhancement magnetic resonance imaging (LGE-MRI) is widely used to assess presence, location and extent of regional scar or fibrotic tissue in the myocardium. T2-weighted CMR images are able to identify edema and acute or recent myocardial ischemic injury, and have been employed to distinguish acute coronary syndrome (ACS) from non-ACS as well as acute from chronic myocardial infarction. On the other hand, balanced - Steady State Free Precession (bSSFP) cine sequence presents clear boundaries for the cardiac anatomical regions, often unclear in the first two modalities due the presence of pathological regions.

LGE and T2-weighted are well-established techniques to many CMR examinations, but there are challenges in their quantification and interpretation due to a variety of factors. First, image analysis depends on image quality which can be affected by CMR acquisition protocol. Suboptimal parameters such as inversion time (TI), repetition time (TR), echo time (TE) need to be correctly identified in order to maximize the difference in intensity curves between pathological and non pathological regions, but also to minimize inter-subject acquisitions variability. Additionally, timing after contrast administration in LGE is important to allow sufficient wash-out of the contrast agent. On top of that, the variability in morphology and texture of infarcted, edemic areas and the combination of both leads to a difficult automation of the process. For this reason, manual and automated techniques with no user interaction for infarct borders detection often results in significant within-patient variability [4,6,10,11].

In order to explore the complementary nature of existing modalities for the purpose of myocardial pathology segmentation, the MyoPS challenge is proposed. It includes a challenging data distribution of 45 multi-modality subjects with the goal of doing an accurate automatic infarcted and edemic regions segmentation.

In this work, we propose a challenge solution based on a stacked BCDU-NET late fusion architecture including localisation and segmentation stages. Additionally, we tackle the insufficient training size by means of state-of-the-art generative adversarial models [5,12]. To do so, we propose an image synthesis strategy based on Semantic Image Synthesis with Spatially-Adaptive Normalization[7]. The results demonstrate that the proposed model improves 1-stage equivalent architectures and benefits from the addition of synthetic anatomically meaningful samples.

2   Materials and Methods

2.1   Dataset

A set of 45 cases of multi-sequence CMR are collected for the challenge. Each case refers to a patient with three CMR sequences, i.e., LGE, T2 and bSSFP CMR. All clinical data have got institutional ethic approval and have been anonymized. The data released have been pre-processed using the MvMM method [13,14] to align the three-sequence CMR into a common space and to resample them into the same spatial resolution.

The provided gold standard labels of interest for the challenge are LV myocardial edema (label 1220) and LV myocardial scars (label 2221). Additional annotations of cardiac structures are provided: left ventricular (LV) blood pool (label 500), right ventricular blood pool (label 600) and LV normal myocardium (label 200). Thus, the evaluation of the test data will be focused on the myocardial pathology segmentation, i.e., scars and edema. The inter-observer variation of manual scar segmentation, in terms of Dice, was 0.5243 ± 0.1578, which gives an insight of the difficulty of the task.

2.2   Proposed Method

An overview of the proposed automated segmentation method is presented in Fig. 1. The approach consists of two stacked segmentation networks. In brief, after preprocessing, we employ a computationally efficient U-Net [8] on the bSSFP CMR to localize the rounded shape of myocardium which includes the LV normal myocardium, LV myocardial edema and scar tissue. Subsequently, the bSSFP, T2-weighted and LGE CMR are cropped using the bounding box of the localized myocardium. Histogram normalization is then applied on the cropped part of imgages. During the second stage, the cropped multi-sequence CMR is passed to a higher capacity model, the BCDU-Net [1], to segment the myocardium scar and edema. The output is finally post-processed based on anatomical constrains to refine output segmentation masks. The individual stages are explained in detail in the following sections.

Preprocessing. Before the training process, all images were cropped so that they had a pixel size of 256×256. Furthermore, all images were normalised between 0 and 1 within the Region Of Interest (ROI) for each independent modality.

Localization Network. The pathological tissue is located within LV blood pool and LV normal myocardium. Therefore, we first employ a network to localize the myocardial ROI, i.e. a binary segmentation, using cine-MRI as the input modality. Cine-MRI was chosen over the other modalities for this task because it is the most accurate for myocardial boundary detection due to its clear structure definition and lack of appearance of pathological regions. This task will reduce

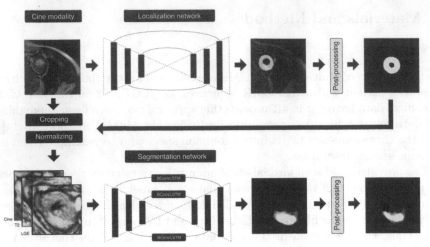

Fig. 1. Overview of the proposed stacked network.

the search space when dealing with scar and edema segmentation by the stacked network. To do that, the myocardium, edema, and scar labels are considered as the foreground, and the other labels (left ventricular blood pool, right ventricular blood pool) as the background. U-Net, [8], is a popular convolutional network architecture for fast and precise segmentation of images which is built upon the Fully Convolutional Network (FCN). The main advantages of this network is that is capable to work well with few training samples, and the network has the potential to make use of the global location and context information at the same time.

This symmetric network is separated in three parts of encoding (contracting), Bottleneck, and decoding (expanding) paths. The encoding path is composed of 4 blocks. In each block we have two 3×3 convolutional layers followed by one 2×2 Max Pooling function and ReLU. In each block, the number of feature maps are doubled, and the size of feature get half. The contracting path aims at progressively capturing context of the input image and increasing the dimension of feature representation block by block. These coarse contextual information are then transferred into the decoding path through skip connections. The output of the last block of the encoder is first passed to the bottleneck which is built by two 3×3 convolutional layers. At the end of bottleneck we have a high dimensional image representation with high semantic information.

The decoding path is composed of four blocks. Each block starts with performing a deconvolution (up-sampling) over the output of previous layer. The corresponding feature maps in the encoding path are then copied to this layer, and are then concatenated with the output of deconvolutional layer. These features are then go through one 3×3 convolutional layers. In each block of the decoder, the size of the feature maps gradually increases and the number of feature maps gradually decreases. The target of decoder in U-Net is to enable precise localisation by using transposed convolutions and recovering the size of the segmentation. Since that data is imbalanced and most of the pixels have background label, we use the weighted binary cross entropy loss to train the network.

In our U-net implementation, for efficiency purposes, the number of classes is used as the number of feature maps in the deconvolutions of the decoding path, as shown in [2,3]. It is also worth mentioning that we do not need a very accurate segmentation result here, since we just crop the smallest bounding box around the myocardium with a small margin of 10 pixels.

Normalisation. The output of the localisation network provides the approximate location of the myocardial region. Therefore, by considering the fact that the myocardial infarcted and edemic regions are within such ROI, we can ignore unwanted background information by finding the smallest bounding box with a small margin around the myocardium. Moreover, an histogram equalisation is applied by modality, avoiding the effect of unuseful background pixels in the pixel histogram redistribution.

Segmentation. We exploit the BCDU-Net [1] to segment the myocardial scar and edema from the normalized myocardium of the three input modalities. The BCDU-Net is an extension of U-Net by including bidirectional convolutional LSTM (BConvLSTM) [9] in the skip connection and reusing feature maps with densely convolutions. The output features of the deconvolutional layer contain more semantic information while the features extracted by the corresponding encoding layer have higher resolution. To combine these two kinds of features, the authors replaced the simple concatenation of the skip connection with nonlinear functions, i.e. BConvLSTM in the BCDU-Net which resulted in more precise segmentation output.

Moreover, the idea of densely connected convolutions is utilized in the bottleneck of the BCDU-Net. By having a sequence of convolutional layers, the network may learn redundant features, therefore, in the bottleneck of the BCDU-Net, features which are learned in each block are passed forward to the next block. The dense blocks help the method to enhance information flow and learn a diverse set of features based on the collective knowledge gained by previous layers. Furthermore, the convergence speed of the network is accelerated by employing Batch Normalization (BN) after the up-convolution filters.

Like U-Net, the encoding path of the BCD-Net includes four steps. Each step consists of two 3×3 convolutional filters followed by a 2×2 max pooling function and ReLU. The depth of feature maps are doubled at each step and the size of each feature map get half. There are two states of BConvLSTM in the skip connection of the BCDU-Net. The second state receives the output of the previous deconvolutional function and the input data of the first one its corresponding feature maps in the encoding path. The output of the second BConvLSTM is then passed to the two 3×3 convolutional filters. Like original U-Net, the decoding path doubles the size of each feature map and halves the number of feature channels layer by layer to reach the original size of the input image after the final layer. To train the network, we use Dice score-based loss.

We propose to combine the three input modalities with a late fusion approach. In other words, the network is trained separately for the three modalities and before the last convolutional layer after the last deconvolutional layer, the three networks are merged.

Implementation Details. All trainings were performed on a NVIDIA 1080 GPU with a batch size of 8. The Adam optimization function with learning rate equal to 1e−4 was used to train both networks. Each network is trained with 50 as the number of epochs. The input size was 256 × 256 for both localization and segmentation networks.

2.3   Data Augmentation Strategy

Online Augmentation. A series of common augmentation techniques were applied to each batched image independently. For the first stacked u-net, these augmentations included random rotations between −15° and 15° and random scaling and offsets of a maximum of 30 pixels. For the second stacked u-net the offset augmentation is avoided due to the fact that images were already center-cropped.

Offline Augmentation. The rationale behind the proposed image synthesis is the insufficient training sample size. Low number of images, variability in modality acquisitions, in location and extent of pathological regions can cause loss of generalisation in CNN-based segmentation algorithms. Thus, in an effort to increase the number of annotated multi-sequence images, semantic image synthesis from annotated mask to multi-sequence CMR is performed in such way that new multi-modality images can be generated from altered versions of real annotations. To achieve this, the Semantic Image Synthesis with Spatially-Adaptive Normalization (SPADE) method [7] was implemented using the PyTorch library provided at this link https://github.com/NVlabs/SPADE. Previous methods [12] directly feed the semantic layout as input to the deep network, which is then processed through stacks of convolution, normalization, and nonlinearity layers. In [7], is shown that this is suboptimal as the normalization layers tend to wash away semantic information, desired for accurate pathology tissue and cardiac structure generation. To address the issue, SPADE uses the input semantic annotation for modulating the activations in normalization layers through a spatially-adaptive, learned transformation. A general overview of the SPADE multi-modality generative model is represented in Fig. 2.

Two SPADE models were generated. For the training/validation subset, a model with 71 training images (17 subjects) was used and 31 validation images (8 subjects) were kept aside. For the final model, all the subjects were used to train an additional SPADE model.

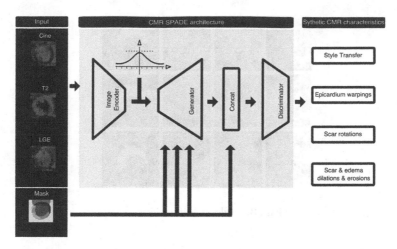

Fig. 2. Overview of the proposed SPADE generative model.

Both models were trained during 45 epochs with a morphological augmentation consisting of warping epicardium contours between pairs of subjects. Both trainings took 24 h on a NVIDIA 1080 GPU with a batch size of 2. The Adam optimizer was used with learning rate of $2 \times 10e-4$, with rst and second moment decay rates of 0 and 0.9, respectively. The Variational Autoencoder (VAE) was generated with a latent dimension of 200.

Once the models were trained, a set of morphological operations were defined in order to generate different versions of real annotations. The resulting anatomical consistent annotations were used then to feed the SPADE models and generate synthetic multi-modality images with controlled characteristics:

Style Transfer. By training the SPADE with a Variational Autoencoder (VAE), the style of the images can be transferred, generating a variety of images with different pathology appearances for the same morphology. The encoder and generator of our SPADE architecture form a VAE, in which the encoder tries to capture the style of the image, while the generator combines the encoded style and the segmentation mask information via the SPADEs to reconstruct the original image. The encoder also serves as a style guidance network at test time to capture the style of target images. For training the VAE, KL-Divergence loss term was used.

Every training image was used to generate a set of latent representations of size 200. The latter were used alone -with random linear combinations and scaling factors- or in conjunction with the methods described below in order to produce the final synthetic multi-modality images. The effect of this technique is shown in Fig. 3, where an original image in first row is transferred to two additional pseudo-random styles, rows 2 and 3.

Epicardium Warpings. As shown in Fig. 4, a set of 8 equidistant landmarks were placed in the epicardial contour of the source and target annotations. Epicar-

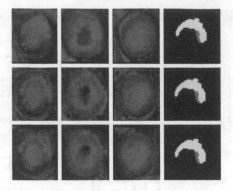

Fig. 3. Style modifications.

dial contours were then warped between pairs of training subjects by means of piecewise affine transformations.

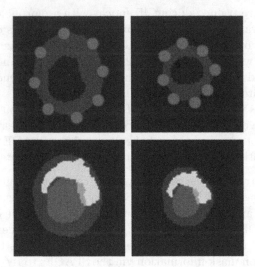

Fig. 4. Epicardial contour warping between a pair of subjects.

Scar and Edema Rotations. As shown in Fig. 7, scar, edema and myocardium labels were combined in a binary mask. The epicardium was then converted to a circular shape, rotated and reconverted to the original shape taking profit of the same technique used in the *Epicardium warpings* section. This set of transformations was then also applied to the original labels, generating a rotated version of the scar and edema within the myocardium. To ensure that the generated segmentations were not too far from the distribution seen by the SPADE generator

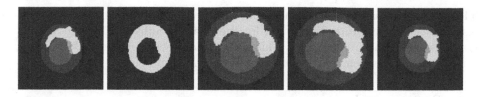

Fig. 5. Morphological operations involved in the scar rotation process.

while covering the label space, the rotation was fixed to four possible values of $[-30°, -20°, 20°, 30°]$ (Fig. 5).

Scar and Edema Dilations and Erosions. A set of random complementary dilations and erosions with a random kernel radius from 1 to 3 pixels were applied to the training annotations. By fixing one of them for the scar label and applying the opposite one for the edema label, we avoid an empty gap between both. Random deletion of edemic labels is also included in this stage. In Fig. 6 shows the effect of an eroded scar and dilated edema.

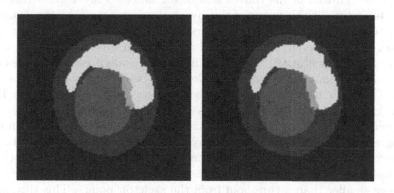

Fig. 6. Morphological operations involved in the scar and edema dilation and erosion process.

Offline Datasets. A group of datasets is generated by means of the augmentation strategies described above. More precisely, for each of the transformable labels, i.e. non-empty annotations, the original images are used up to three times to keep the training size relatively small. This methodology leads to the creation of a set of four datasets, one per type of augmentation, i.e. style transfer alone, pathology rotations, epicardial warping and pathology dilation/erosion. It should be noted that the resulting datasets contain the same amount of real and synthetic data. Additionally, for all datasets, random style transfers are applied after the annotation manipulation in the synthesis stage. In total, each dataset contains 415 images. A fifth dataset is generated by combining all individual four datasets.

This dataset consists of 1660 images and is used to train and validate the models. The same procedure is repeated for the final ensemblea using the SPADE trained over all the training data. This leads to datasets of 597 and 2388 images, for the partial augmentations and the addition, respectively.

2.4 Post-processing

The myocardium, scar and edema-scar segmentations produced from the stacked networks were morphologically processed to satisfy certain anatomical constraints. In short axis CMR, the shape of the myocardium closely resembles that of a ring throughout the apex-base slices. Therefore, slices for which the automatically segmented myocardium is a partial ring must be detected and corrected. To this end, the skeleton of the myocardium was calculated for each slice. Subsequently, spur skeleton branches, i.e. branches consisting of pixels with only one neighboring pixel, were iteratively pruned. For non-complete rings, iterative pruning results in the removal of the entire skeleton. In such cases, the missing arc of the partial ring was completed by adding a circular ring whose thickness is equal to the maximum thickness of the detected myocardium. To construct the ring, the centroid of the convex hull of the detected myocardial region was used as its center. The thickness of the myocardium was given by the distance of the skeleton points to the closest non-myocardial pixel and the maximum among all points was considered. The corrected myocardium was subsequently used to refine the scar segmentation, while an additional step was necessary in the case of the edema-scar region. More precisely, edema can be noticed in the myocardium, but also in the LV blood pool close to the border with the myocardium. Therefore, an extended myocardial mask was created, which contained neighboring LV regions where edema could be localized. In order to achieve this, an artificial ring was constructed by using the myocardium skeleton and the distance of every pixel to it. Pixels belonging to the myocardium or the region enclosed by it were considered to belong to the extended myocardial mask if they were within a distance smaller than a threshold from the skeleton points. This threshold is defined as the maximum myocardium thickness plus a small margin of 6 pixels to account for errors in the myocardium segmentation.

As a first step in the process of refining the scar tissue, 3D components smaller than 100 voxels were considered to be artifacts and were, therefore, excluded from the segmentation mask. Despite good localization of the scar region by the network, we observed a tendency to underestimate the scar region and to produce multiple disconnected components instead of one continuous region. To tackle this issue, the components were connected by using their convex hull in cases where the output of the network consisted of more than one connected components. The area of the convex hull inside an eroded version of the extended myocardium was eliminated. For the erosion, a disk element with radius equal to 20% of the maximum myocardium radius was used. Furthermore, morphological closing of the image with a disk object of radius equal to 90% of the myocardium maximum thickness was performed to enlarge the component's border without losing the form of the original shape boundary in cases where only one component was observed. Lastly, areas outside the corrected myocardium and the joined edema-scar mask regions were excluded from the final scar segmentation.

In the case of the refinement of the joined edema-scar mask, 3D components of size smaller than 300 voxels were considered as artifacts. In addition, regions of edema-scar outside the extended myocardial area were excluded from the final segmentation by performing element-wise multiplication of the artificial extended myocardium region mask with the edema-scar segmentation.

3 Results

3.1 Protocol and Metrics of the Challenge

In order to train our models and generate the ablation study, the training set is divided in two partitions. From the original 25 subjects, 8 of them are kept aside for validation, with the aim of preserving a large pool of subjects in the validation stage. The decision is motivated by the variability in image quality and the presence of difficult cases that may lead to a sub-optimal model selection. Moreover, this allows us to have a sufficient validation size to evaluate the post-processing algorithm. For the same reason, we avoided to preserve a test partition that leads to a conflict between validation and testing results and generates additional uncertainty when selecting the best method. After model generation, selection, evaluation and post-processing, 3D Dice scores are computed to select the final models taking into consideration the post-processing gains. For all the experiments, 2D Dice score is used as objective loss function, except for the localisation U-net, where the selected loss is binary weighted cross-entropy.

3.2 Ablation Study

We performed a detailed ablation study in order to quantify the effect of every component of the proposed methodology individually. The results in terms of 2D Dice score (mean ± standard deviation), which is the accuracy evaluation metric used in the loss function of this work, are summarized in Table 1. In brief, our first experiment involved segmenting the scar and scar + segmentation using solely the original data without performing inter-stage normalization or offline augmentation. This resulted in a Dice score equal to 0.202 ± 0.286 and 0.170 ± 0.253 for scar and scar + edema, respectively. The low accuracy demonstrates the extremely challenging nature of the task and the need for incorporating a ROI-based normalization between stages and novel augmentation strategies. To test our assumption, we added the inter-stage cropping and normalization step to enhance the contrast between scar and edema and the rest of the tissue within the myocardial ROI where the pathological tissue is expected to localized. The mean dice score increased by 24.70% for scar and 33.80% for scar + edema.

We then compared the improvement offered by any of the four types of offline augmentation, i.e. style transfer alone, pathology rotations, epicardial warping and pathology dilation/erosion. Style transfer produced an improvement in terms

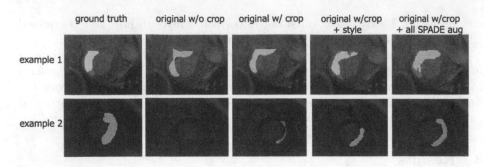

Fig. 7. Segmentation examples combining different sets of training data, showing the improvement of SPADE synthesis.

of Dice by 9% and 14.4% for scar and scar + edema, respectively. The effects of epicardium warping and scar and edema rotation, were lower than that of style-transfer, but yet non-negligable. More precisely, the mean dice increased by 4.1% for scar and 7.8% for scar + edema in the case of epicardium warping. Similarly, when scar and edema rotation were applied the offered improvement was 1.7% for scar and 4.6% for scar + edema. Interestingly, scar and edema dilation and erosion did not provide any significant improvement in the scar tissue, but offered a 10.4% mean improvement in Dice for the scar + edema region. Subsequently, we combined the four types of data-augmentation. We observed a Dice score of 0.518 ± 0.286 and 0.617 ± 0.253 for scar and scar + edema, respectively. This indicates that for the case of pathological tissue segmentation the most effective augmentation type is style transfer, while morphological augmentations have a more limited effect. We speculate that this might be related to the highly irregular shape of the pathological tissue. However, these types of morphological augmentations might be important in other more regular structures. In this work, to account for possible variability found in the test sample non present in the training set, for the final model, we decided to use the combination of all augmentation types, presented as "All spade" in Table 1. Nonetheless, future work will focus on using the style transfer only for pathological tissue segmentation.

Lastly, we evaluated the improvement offered by applying post-processing on the outputs of the localization and segmentation networks. A visual example of the improvement can be seen in Fig. 8. Post-processing produces a continuous scar region, while both edema and scar after post-processing are localized within the myocardial area and in the close vicinity of left ventricle, as physiologically expected.

3.3   Challenge Results

In order to obtain the final predictions, two ensembles are generated. For the first ensemble, a set of 5 models is generated with 10 consecutive training samples and 5 consecutive validation samples, with a roll factor of 5. For the second ensemble,

Table 1. 2D Dice score (mean ± standard deviation) of the proposed method for scar and scar + edema for different data.

Data	Scar	Scar + Edema
Original data	0.202 ± 0.286	0.170 ± 0.253
Original data + cropping and normalizing	0.449 ± 0.261	0.508 ± 0.243
Style transfer	0.548 ± 0.250	0.640 ± 0.192
Epicardium warping	0.490 ± 0.260	0.586 ± 0.222
Scar and edema rotation	0.466 ± 0.241	0.554 ± 0.224
Scar and edema dilation and erosion	0.458 ± 0.299	0.600 ± 0.224
All spade	0.518 ± 0.286	0.617 ± 0.253

a set of 15 models is generated with 22 consecutive training samples and 3 consecutive validation subjects, with a roll factor of 2, making the validation set to share one subject between consecutive models in the case of the 15 models ensemble.

The confidence maps of each one of the 5 models are averaged together. The final predictions of the 20 unseen test subjects provided by the challenge organization are defined as the maximum average probability of each pixel belonging to each class, maximizing the expected results and reducing the variance. The same procedure was applied to the 15 models ensemble. After that, post-processing, as described in Sect. 2.4, is applied to further enhance the model's output. The effect of the ensemble size can be observed in Table 2. The bigger ensemble obtained better results due to the bigger training sizes. The effect of the low validation size was noticeable as a noisier validation curve, and attenuated by means of a greater regularisation power, with an overall improved accuracy. The quantitative effect of post-processing is also appreciated. The 15 models ensemble captured a greater number of non-trivial unconnected components. In combination with the convex hull process described in Sect. 2.4, for the 15 models ensemble the post-processing generated an improvement in accuracy of 2.9% for scar and 1.1% for scar + edema, respectively.

Table 2. 3D Dice score for the final testing set of 20 subjects.

Data	Scar	Scar + Edema
5 models ensemble	0.625 ± 0.255	0.677 ± 0.146
5 models ensemble + post-processing	0.635 ± 0.281	0.692 ± 0.143
15 models ensemble	0.636 ± 0.243	0.687 ± 0.131
15 models ensemble + post-processing	0.665 ± 0.241	0.698 ± 0.128

4   Discussion

This work proposes a novel approach to address automatic multi-sequence CMR pathology segmentation. The method is based on a two-staged process and leverages advanced state-of-the-art deep learning techniques. CMR pathology segmentation is a particularly challenging task even for the expert clinician due to the large variability in imaging quality and morphology of pathological regions. To tackle this limitation, we focus on reducing the task complexity. To this end, a localisation U-net is used to localize the myocardial ROI. Subsequently, the detected ROI is used to partially address the problem of intra- and inter-subject variability in signal intensity by using the bounding box of the ROI to crop the CMR images and perform a refined normalisation within the cropped region. The normalised CMR are then fed to a BCDU-net in order to perform the pathologic tissue segmentation. BCDU-net effectiveness has been previously demonstrated and is related to the bidirectional flow of the gradient. In addition, we address the problem of insufficient training examples by means of multi-modality semantic

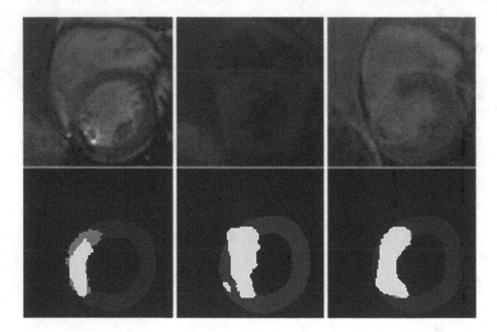

Fig. 8. Improvement offered by applying post-processing on the outputs of the localization and segmentation networks. On the top row, a slice from the bSSFP (left), T2-weighted (middle) and LGE (right) CMR are provided for one subject of the training dataset used as validation subject during training. On the bottom row, the corresponding manual segmentations for myocardium, scar and edema (left), the combined output of the two networks before (middle) and after (right) post-processing are provided. Post-processing permits to connect the two disconnected components produced by the network and constrain the segmentation within the myocardial area and neighboring LV area.

image synthesis using morphological and style transformations. This approach increases the variability of the training samples in terms of the location of the infarcted and edemic tissues within the myocardium, as well as, in terms of their appearance. The validation shows the effect of the stacked architecture with inter-stage normalisation, giving an insight about the importance of standarisation for multi-modality medical imaging acquisitions. Moreover, consistent results across the different semantic manipulations and their respective synthesis, indicate the potential of this set of transformations for enriching and improving generalization of multi-modality cardiac pathology segmentation algorithms. Future work includes the implementation of an end-to-end model as well as the exploration of the generated synthetic data in detail with the aim of enhancing interpretability and quality of the image synthesis methods.

Acknowledgement. This work was partly funded by the European Union's Horizon 2020 research and innovation programme under grant agreement no. 825903 (euCanSHare project). This work has been partially supported by the Spanish project PID2019-105093GB-I00 (MINECO/FEDER, UE) and CERCA Programme/Generalitat de Catalunya.). This work is partially supported by ICREA under the ICREA Academia programme. KL is supported by the Ramon y Cajal Program of the Spanish Ministry of Economy and Competitiveness under grant no. RYC-2015-17183.

References

1. Azad, R., Asadi-Aghbolaghi, M., Fathy, M., Escalera, S.: Bi-directional ConvLSTM U-Net with densley connected convolutions. In: 2019 IEEE/CVF International Conference on Computer Vision Workshop (ICCVW). IEEE, October 2019. https://doi.org/10.1109/iccvw.2019.00052
2. Baumgartner, C.F., Koch, L.M., Pollefeys, M., Konukoglu, E.: An exploration of 2D and 3D deep learning techniques for cardiac MR image segmentation. In: Pop, M., et al. (eds.) STACOM 2017. LNCS, vol. 10663, pp. 111–119. Springer, Cham (2018). https://doi.org/10.1007/978-3-319-75541-0_12
3. Campello, V.M., Martín-Isla, C., Izquierdo, C., Petersen, S.E., Ballester, M.A.G., Lekadir, K.: Combining multi-sequence and synthetic images for improved segmentation of late gadolinium enhancement cardiac MRI. In: Pop, M., et al. (eds.) STACOM 2019. LNCS, vol. 12009, pp. 290–299. Springer, Cham (2020). https://doi.org/10.1007/978-3-030-39074-7_31
4. Flett, A.S., et al.: Evaluation of techniques for the quantification of myocardial scar of differing etiology using cardiac magnetic resonance. JACC Cardiovasc. Imaging 4(2), 150–156 (2011). https://doi.org/10.1016/j.jcmg.2010.11.015
5. Goodfellow, I., et al.: Generative adversarial nets. In: Advances in Neural Information Processing Systems, pp. 2672–2680 (2014)
6. Klem, I., et al.: Sources of variability in quantification of cardiovascular magnetic resonance infarct size - reproducibility among three core laboratories. J. Cardiovasc. Magn. Reson. **19**(1) (2017). https://doi.org/10.1186/s12968-017-0378-y
7. Park, T., Liu, M.Y., Wang, T.C., Zhu, J.Y.: Semantic image synthesis with spatially-adaptive normalization. In: 2019 IEEE/CVF Conference on Computer Vision and Pattern Recognition (CVPR). IEEE, June 2019. https://doi.org/10.1109/cvpr.2019.00244

8. Ronneberger, O., Fischer, P., Brox, T.: U-Net: convolutional networks for biomedical image segmentation. In: Navab, N., Hornegger, J., Wells, W.M., Frangi, A.F. (eds.) MICCAI 2015. LNCS, vol. 9351, pp. 234–241. Springer, Cham (2015). https://doi.org/10.1007/978-3-319-24574-4_28

9. Song, H., Wang, W., Zhao, S., Shen, J., Lam, K.-M.: Pyramid dilated deeper ConvLSTM for video salient object detection. In: Ferrari, V., Hebert, M., Sminchisescu, C., Weiss, Y. (eds.) ECCV 2018. LNCS, vol. 11215, pp. 744–760. Springer, Cham (2018). https://doi.org/10.1007/978-3-030-01252-6_44

10. Tao, Q., Piers, S.R., Lamb, H.J., van der Geest, R.J.: Automated left ventricle segmentation in late gadolinium-enhanced MRI for objective myocardial scar assessment. J. Magn. Reson. Imaging **42**(2), 390–399 (2014). https://doi.org/10.1002/jmri.24804

11. Thiele, H., Kappl, M.J., Conradi, S., Niebauer, J., Hambrecht, R., Schuler, G.: Reproducibility of chronic and acute infarct size measurement by delayed enhancement-magnetic resonance imaging. J. Am. Coll. Cardiol. **47**(8), 1641–1645 (2006). https://doi.org/10.1016/j.jacc.2005.11.065

12. Wang, T.C., Liu, M.Y., Zhu, J.Y., Tao, A., Kautz, J., Catanzaro, B.: High-resolution image synthesis and semantic manipulation with conditional GANs. In: 2018 IEEE/CVF Conference on Computer Vision and Pattern Recognition. IEEE, June 2018. https://doi.org/10.1109/cvpr.2018.00917

13. Zhuang, X.: Multivariate mixture model for cardiac segmentation from multi-sequence MRI. In: Ourselin, S., Joskowicz, L., Sabuncu, M.R., Unal, G., Wells, W. (eds.) MICCAI 2016. LNCS, vol. 9901, pp. 581–588. Springer, Cham (2016). https://doi.org/10.1007/978-3-319-46723-8_67

14. Zhuang, X.: Multivariate mixture model for myocardial segmentation combining multi-source images. IEEE Trans. Pattern Anal. Mach. Intell. **41**(12), 2933–2946 (2019). https://doi.org/10.1109/tpami.2018.2869576

EfficientSeg: A Simple But Efficient Solution to Myocardial Pathology Segmentation Challenge

Jianpeng Zhang[1], Yutong Xie[1], Zhibin Liao[2,3], Johan Verjans[2,3], and Yong Xia[1(✉)]

[1] National Engineering Laboratory for Integrated Aero-Space-Ground-Ocean Big Data Application Technology, School of Computer Science and Engineering, Northwestern Polytechnical University, Xi'an 710072, China
yxia@nwpu.edu.cn
[2] The University of Adelaide, Adelaide, Australia
[3] South Australian Health and Medical Research Institute, Adelaide, Australia

Abstract. Myocardial pathology segmentation is an essential but challenging task in the computer-aided diagnosis of myocardial infraction. Although deep convolutional neural networks (DCNNs) have achieved remarkable success in medical image segmentation, accurate segmentation of myocardial pathology remains challenging, due to the low soft-tissue contrast, irregularity of pathological targets, and limited training data. In this paper, we propose a simple but efficient DCNN model called *EfficientSeg* to segment the regions of edema and scar in multi-sequence cardiac magnetic resonance (CMR) data. In this model, the encoder uses EfficientNet as its backbone for feature extraction, and the decoder employs a weighted bi-directional feature pyramid network (BiFPN) to predict the segmentation mask. The former has a much improved image representation ability but with less computation cost than traditional convolutional networks, while the latter allows easy and fast multi-scale feature fusion. The loss function of EfficientSeg is defined as the combination of Dice loss, cross entropy loss, and boundary loss. We evaluated EfficientSeg on the Myocardial Pathology Segmentation (MyoPS 2020) Challenge dataset and achieved a Dice score of 64.71% for scar segmentation and a Dice score of 70.87% for joint edema and scar segmentation. Our results indicate the effectiveness of the proposed EfficientSeg model for myocardial pathology segmentation.

Keywords: Myocardial pathology segmentation · Deep learning · Cardiac magnetic resonance imaging

1 Introduction

Automated myocardial pathology segmentation using cardiac magnetic resonance (CMR) imaging is able to assist doctors in the accurate assessment of

© Springer Nature Switzerland AG 2020
X. Zhuang and L. Li (Eds.): MyoPS 2020, LNCS 12554, pp. 17–25, 2020.
https://doi.org/10.1007/978-3-030-65651-5_2

myocardial viability and fast evaluation of myocardial infarction development, playing an essential role in the diagnosis and treatment management for the patients suffering from myocardial infarction. However, this task remains challenging due to two reasons: (1) myocardial pathology regions, *e.g.*, scars and edema, vary significantly in the visual appearance; and (2) the borders between pathology regions and surrounding normal organs or tissues appear blurry and ambiguous.

Traditional segmentation methods for CMR data mainly include the prior knowledge based methods [8,11], graph-cuts [1], and atlas-based registration [8]. Recently, a number of deep convolutional neural networks (DCNNs) [2,4,10,12,17,19] have been proposed for CMR data segmentation, which achieve superior performance over those traditional methods. Dou *et al.* [4] introduced the knowledge distillation to DCNNs, aiming to employ the cardiac computed tomography data to assist CMR data segmentation. Liu *et al.* [10] proposed the Pseudo-3D CNN, which aims to reduce the size of the network while preserving the spatial structure information in 3D data. Wang *et al.* [17] modified the popular U-Net by incorporating the squeeze-and-excitation residual module and selective kernel module into the down-sampling and up-sampling stages, respectively. Despite their improved performance, most of these DCNNs focus on the segmentation of the ventricle, ventricle myocardium, atrium blood cavity, ventricle blood cavity or ascending aorta from CMR data. Little effort has been devoted to the fully automated myocardial pathology segmentation. To this end, the MICCAI 2020 has launched the Myocardial Pathology Segmentation Combining Multi-sequence CMR (MyoPS 2020) Challenge to accelerate the technical development and clinical deployment of automated CMR segmentation solutions.

In this paper, we propose a simple but efficient DCNN model called *EfficientSeg* for accurate myocardial pathology segmentation in multi-sequence CMR scans. In this model, we use EfficientNets, which is a recent convolutional neural network with a strengthened ability to representation learning [15], as the encoder backbone for feature extraction, and also incorporate the bi-directional feature pyramid network (BiFPN) [16] into the decoder for effective multi-scale feature fusion. To address the issue caused by low soft tissue contrast, we train our EfficientSeg model via minimizing a weighted sum of the Dice loss, cross entropy loss, and boundary loss. Among them, the boundary loss helps the model focus on the myocardial pathology boundaries, and thus also alleviates the class-imbalanced problem. We have evaluated the proposed EffcientSeg model on the MyoPS 2020 Challenge dataset and achieved a Dice score of 64.71% for scar segmentation and a Dice score of 70.87% for joint edema and scar segmentation.

2   Dataset

The dataset used for this study was provided by the MyoPS 2020 Challenge, which aims to develop automated and accurate algorithms to segment the myocardial pathology, including the left ventricular myocardial scars and edema.

The challenge dataset consists of 45 cases with multi-sequence CMR scans, each case containing a late gadolinium enhancement (LGE) CMR sequence, a T2-weighted CMR sequence, and a balanced Steady State Free Precession (bSSFP) cine sequence. All cases have been aligned into the common space and resampled into the same spatial resolution by using the MvMM method [20,21]. The dataset was officially split into a training set of 25 cases and a testing set of 20 cases. The voxel-wise annotations of training cases are publicly available, while the annotations of testing cases are withheld for online evaluation.

3   Method

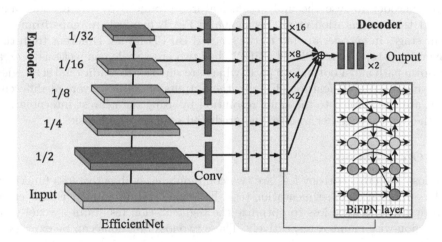

Fig. 1. Diagram of the proposed EfficientSeg model. "Conv": convolution layer; ×2: double the feature resolution using nearest interpolation; ⊕: element-wise summation.

Our proposed EfficientSeg model has an encoder-and-decoder architecture, in which the encoder backbone is the state-of-the-art EfficientNets [15] and the decoder is constructed based on BiFPN [16].

3.1   Encoder

Scaling up convolutional neural networks is an important way to improve the representation ability and thus to obtain better accuracy. Tan *et al.* [15] developed a strong baseline model, EfficientNet-B0, by using neural architecture searching. The obtained Efficient-B0 is mainly composed of mobile inverted bottleneck blocks [13,14] with the squeeze-and-excitation attention mechanism [5]. Based on the baseline model, they proposed a family of EfficientNets using a simple yet effective compound scaling method, and achieved the state-of-the-art performance on several imageg classification benchmarks. For an improved representation ability, we employ the ImageNet-pretrained EfficientNets as feature extractor in the encoder, aiming to extract strong features from CMR sequences. According to the resources available for model scaling, EfficientNets contain eight networks with different scales, varying from B0 to B7. To balance the accuracy and computational complexity, we choose EfficientNet-B1, EfficientNet-B2, and EfficientNet-B3 for the encoder, respectively.

3.2   Decoder

BiFPN contains an efficient multi-scale feature fusion strategy, which achieves the state-of-the-art performance on object detection tasks [16]. Considering the objects (*i.e.* scars and edema) with different shapes and sizes, we employ BiFPN as the decoder to fuse the multi-scale features produced by the encoder and predict the segmentation mask. As shown in Fig. 1, five feature maps from different stages in encoder are fed to the stacked BiFPN layers. Different from the conventional top-down FPN [9], BiFPN has two cross-scale connections, *i.e.*, a top-down path and a bottom-up path, which are superior to traditional strategies for fusing multi-scale features. Finally, the outputs of BiFPN layers at different scales are up-sampled to the same resolution by using the nearest interpolation and then added together to form the predicted segmentation mask.

3.3   Optimization

Dice loss and cross entropy loss are two commonly used objective loss functions in the optimization of segmentation task. Following [18], we jointly use the cross entropy loss and Dice loss to optimize the segmentation results in a voxel-wise and region-wise manner, respectively. These two loss functions can be expressed as

$$\mathfrak{L}_{CE} = -\frac{1}{N}\sum_i^N\sum_c^C Y_i^c \log P_i^c \tag{1}$$

$$\mathfrak{L}_{Dice} = 1 - \frac{1}{C}\sum_c^C \frac{2\sum_i^N Y_i^c P_i^c}{\sum_i^N Y_i^c + \sum_i^N P_i^c} \tag{2}$$

where C is the number of categories, N is the number of voxels, P_i^c represents the predicted probability of voxel i belonging to class c, and Y_i^c represents the ground truth label of voxel i.

Besides, we observe the low contrast between the myocardial pathology regions and surrounding tissues. Considering the poor discrimination around boundaries, we also employ the boundary loss proposed in [7] to enforce the model to pay more attention to boundary regions. Let ∂Y represent the boundary of Y, and Ω represent the spatial domain of the given image. A signed distance between a voxel $x \in \Omega$ and its nearest voxel $z_{\partial Y}(x)$ on contour ∂Y can be defined as

$$\phi_Y(x) = \begin{cases} -||x - z_{\partial Y}(x)||, & if \ x \in Y \\ ||x - z_{\partial Y}(x)||, & otherwise \end{cases} \tag{3}$$

Then, the boundary loss can be formulated as

$$\mathcal{L}_{Bound} = \int_{\Omega} \phi_Y(x)P(x)dx \qquad (4)$$

The proposed EfficientSeg model uses the following combined loss

$$\mathcal{L} = \alpha(\mathcal{L}_{CE} + \mathcal{L}_{Dice}) + (1-\alpha)\mathcal{L}_{Bound} \qquad (5)$$

where $\alpha = 1 - \frac{1-0.01}{K} \cdot k$ is a weighting factor, K is the total number of epochs and k is the index of current epoch. At the beginning of training process, cross entropy loss and Dice loss play the principal role in the optimization. With the training going on, the boundary loss exerts more influence on the optimization.

3.4   Emplementation Details

Since the dynamic ranges of different CMR sequences in different cases are variable, it is necessary to normalize the value of each sequence to avoid the initial bias of the segmentation process [6]. For each CMR sequence, we first normalize the voxel values to zero mean and unit standard deviation, and then concatenate three CMR sequences into a multi-sequence volume with three channels.

Due to the large inter-slice spacing, the segmentation is performed on a slice-by-slice basis. On each slice of a training case, we randomly sample the patches of size 288×288 to train the EfficientSeg model. To accelerate the training procedure and enlarge the batch size, we utilize the mixed precision training strategy[1]. According to the varying scale of backbone series, we set the batch size to 64, 48, and 32 for EfficientSeg-B1, EfficientSeg-B2, and EfficientSeg-B3, respectively. We use the Adam algorithm with an initial learning rate of 1×10^{-4} to optimize the EfficientSeg model. The learning rate was dynamically decayed by a factor of 5 when the validation loss was not improved any more in the last 30 epochs. We set the maximum epoch to 500. To alleviate the potential overfitting, we employ some simple online data augmentation techniques, including randomly cropping and flipping along spatial dimensions.

During the testing stage, we perform the test time augmentation by mirror flipping the input along two axes and then averaging the outputs of two flipped inputs and the original input. To reduce the false positive predictions, we adopt a simple post-processing strategy to remove small isolated predicted regions. We calculate the connected regions of each category in each slice. For the predicted regions of scar, we remove the small regions with less than N_1 voxels and replace them with the surrounding labels. For the predicted regions of edema and scars, we remove the small regions with less than N_2 voxels and replace them with the surrounding labels. Here, we set N_1 to 60 and N_2 to 200 according to the performance observations on the validation set.

Following the rule of the MyoPS 2020 Challenge, we use the Dice score as the evaluation metric for scars and edema segmentation.

---

[1] https://nvidia.github.io/apex.

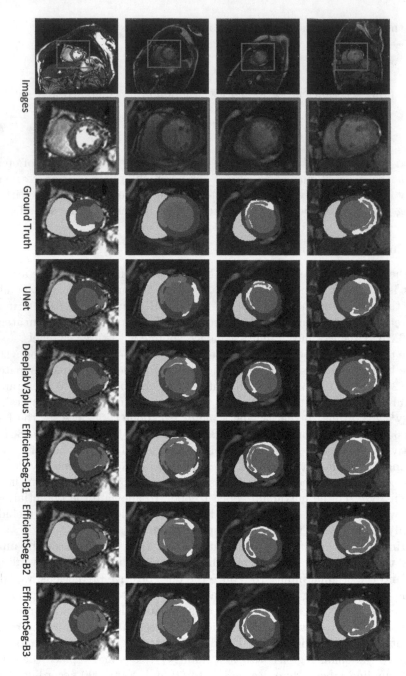

Fig. 2. Visualization of segmentation results. Colors represent the different regions: left ventricular normal myocardium; left ventricular blood pool; right ventricular blood pool; ; left ventricular myocardial scars. Note that the provided gold standard labels include five categories, while the evaluation of the test data only focuses on the myocardial pathology segmentation, i.e., scars and edema.

Table 1. Comparison of different segmentation methods on the test set. †: Fine-tuning the model using the pretrained ImageNet weights [3]; ⋆: Final submission to the MyoPS 2020 Challenge; Ensembles: Ensemble B1, B2 and B3; pp: post-processing.

Method	# Parameters ($\times 10^6$)	Dice score %	
		Scar	Edema+Scar
U-Net	17.3	59.15	62.56
DeeplabV3plus	54.7	60.60	64.71
DeeplabV3plus †	54.7	63.01	69.05
EfficientSeg-B1 †	6.5	62.94	68.98
EfficientSeg-B2 †	7.9	63.08	68.78
EfficientSeg-B3 †	11.7	62.56	69.47
EfficientSeg-Ensembles †	/	64.35	70.10
EfficientSeg-Ensembles + pp † ⋆	/	**64.71**	**70.87**

4 Results

In Table 1, we compared the segmentation performance of different methods, including U-Net, DeeplabV3plus, and the proposed EfficientSeg with different backbones, on the test set. For a fair comparison, all competing methods and our method use the same training and testing strategy. It reveals that (1) fine-tuning from the pretrained weights (DeeplabV3plus and EfficientSeg w/pretraining) achieves better performance than training from scratch (U-Net and DeeplabV3plus w/o pretraining); (2) our EfficientSeg is able to obtain comparable (even better) performance with less parameters than DeeplabV3plus; (3) our ensemable strategy has a positive effect on the performance; and (4) our proposed EfficientSeg model achieves the 64.71% Dice score for scar segmentation and 70.87% Dice score for joint edema and scar segmentation using the ensemble strategy (*i.e.*, averaging outputs of multiple models) and post-processing. In Fig. 2, we visualized the segmentation results obtained by U-Net, DeeplabV3plus and our EfficientSeg-B1/B2/B3 on the validation set.

To demonstrate the effectiveness of our combined segmentation loss, we attempt to train the EfficientSeg-B1 model with different loss functions, including the cross entropy loss $\mathcal{L}_{CE}$, Dice loss $\mathcal{L}_{Dice}$, the sum of cross-entropy loss and Dice loss $\mathcal{L}_{CE} + \mathcal{L}_{Dice}$, and the sum of cross-entropy loss, Dice loss and boundary loss $\mathcal{L}_{CE} + \mathcal{L}_{Dice} + \mathcal{L}_{Bound}$. The results in Table 2 reveals that, although using the cross-entropy loss alone results in a good performance on the segmentation of Edema and Scar, adding the Dice loss and boundary loss one by one to the segmentation loss improves the performance continuously. Meanwhile, the superior performance of our combined loss over the combination of cross-entropy loss and Dice loss confirms the effectiveness of using the boundary loss to impose constraints on boundary regions.

Table 2. Comparison of different loss functions on the test set.

Method	Dice score %	
	Scar	Edema+Scar
EfficientSeg-B1 + $\mathcal{L}_{CE}$	61.59	67.61
EfficientSeg-B1 + $\mathcal{L}_{Dice}$	62.13	68.21
EfficientSeg-B1 + $\mathcal{L}_{CE}$ + $\mathcal{L}_{Dice}$	62.41	68.11
EfficientSeg-B1 + $\mathcal{L}_{CE}$ + $\mathcal{L}_{Dice}$ + $\mathcal{L}_{Bound}$ (Ours)	**62.94**	**68.98**

5 Conclusion

In this paper, we propose a simple but efficient model, *i.e.*, EfficientSeg, to address the myocardial pathology segmentation challenge. We choose the state-of-the-art EfficientNet to extract the representative features from multi-sequence CMR data, and then use BiFPN to fuse multi-scale feature maps for the accurate myocardial pathology segmentation. Besides, we optimize the EfficientSeg model via minimizing the combination of voxel-wise cross entropy loss, region-wise Dice loss, and boundary loss. On the online testing, the proposed EfficientSeg model achieves a Dice score of 64.71% for scar segmentation, and a Dice score of 70.87% for joint edema and scar segmentation.

Acknowledgment. Y. Xie, J. Zhang, and Y. Xia were supported by the National Natural Science Foundation of China under Grants 61771397. Y. Xie was supported by the Innovation Foundation for Doctor Dissertation of Northwestern Polytechnical University under Grants CX202010. We appreciate the organizers of the MyoPS 2020 Challenge for their efforts devoted to collect and share the CMR datasets for the development and evaluation of automated myocardial pathology segmentation algorithms.

References

1. Alba, X., Figueras i Ventura, R.M., Lekadir, K., Tobon-Gomez, C., Hoogendoorn, C., Frangi, A.F.: Automatic cardiac LV segmentation in MRI using modified graph cuts with smoothness and interslice constraints. Magn. Reson. Med. **72**(6), 1775–1784 (2014)
2. Chen, C., et al.: Unsupervised multi-modal style transfer for cardiac MR segmentation. In: Pop, M., et al. (eds.) STACOM 2019. LNCS, vol. 12009, pp. 209–219. Springer, Cham (2020). https://doi.org/10.1007/978-3-030-39074-7_22
3. Deng, J., et al.: ImageNet: a large-scale hierarchical image database. In: 2009 IEEE Conference on Computer Vision and Pattern Recognition, pp. 248–255. IEEE (2009)
4. Dou, Q., Liu, Q., Heng, P.A., Glocker, B.: Unpaired multi-modal segmentation via knowledge distillation. arXiv preprint arXiv:2001.03111 (2020)
5. Hu, J., Shen, L., Sun, G.: Squeeze-and-excitation networks. In: Proceedings of the IEEE Conference on Computer Vision and Pattern Recognition, pp. 7132–7141 (2018)

6. Isensee, F., Kickingereder, P., Wick, W., Bendszus, M., Maier-Hein, K.H.: No new-net. In: Crimi, A., Bakas, S., Kuijf, H., Keyvan, F., Reyes, M., van Walsum, T. (eds.) BrainLes 2018. LNCS, vol. 11384, pp. 234–244. Springer, Cham (2019). https://doi.org/10.1007/978-3-030-11726-9_21

7. Kervadec, H., Bouchtiba, J., Desrosiers, C., Granger, E., Dolz, J., Ayed, I.B.: Boundary loss for highly unbalanced segmentation. In: International Conference on Medical Imaging with Deep Learning, pp. 285–296 (2019)

8. Kurzendorfer, T., Forman, C., Schmidt, M., Tillmanns, C., Maier, A., Brost, A.: Fully automatic segmentation of left ventricular anatomy in 3D LGE-MRI. Comput. Med. Imaging Graph. **59**, 13–27 (2017)

9. Lin, T.Y., Dollár, P., Girshick, R., He, K., Hariharan, B., Belongie, S.: Feature pyramid networks for object detection. In: Proceedings of the IEEE Conference on Computer Vision and Pattern Recognition, pp. 2117–2125 (2017)

10. Liu, T., et al.: Pseudo-3D network for multi-sequence cardiac MR segmentation. In: Pop, M., et al. (eds.) STACOM 2019. LNCS, vol. 12009, pp. 237–245. Springer, Cham (2020). https://doi.org/10.1007/978-3-030-39074-7_25

11. Lu, Y., Wright, G., Radau, P.E.: Automatic myocardium segmentation of LGE MRI by deformable models with prior shape data. J. Cardiovasc. Magn. Reson. **15**(1), 1–2 (2013)

12. Ronneberger, O., Fischer, P., Brox, T.: U-Net: convolutional networks for biomedical image segmentation. In: Navab, N., Hornegger, J., Wells, W.M., Frangi, A.F. (eds.) MICCAI 2015. LNCS, vol. 9351, pp. 234–241. Springer, Cham (2015). https://doi.org/10.1007/978-3-319-24574-4_28

13. Sandler, M., Howard, A., Zhu, M., Zhmoginov, A., Chen, L.C.: Mobilenetv 2: inverted residuals and linear bottlenecks. In: Proceedings of the IEEE Conference on Computer Vision and Pattern Recognition, pp. 4510–4520 (2018)

14. Tan, M., et al.: MnasNet: platform-aware neural architecture search for mobile. In: Proceedings of the IEEE Conference on Computer Vision and Pattern Recognition, pp. 2820–2828 (2019)

15. Tan, M., Le, Q.V.: EfficientNet: rethinking model scaling for convolutional neural networks. arXiv preprint arXiv:1905.11946 (2019)

16. Tan, M., Pang, R., Le, Q.V.: EfficientDet: scalable and efficient object detection. In: Proceedings of the IEEE Conference on Computer Vision and Pattern Recognition, pp. 10781–10790 (2020)

17. Wang, X., et al.: SK-Unet: an improved U-Net model with selective kernel for the segmentation of multi-sequence cardiac MR. In: Pop, M., et al. (eds.) STACOM 2019. LNCS, vol. 12009, pp. 246–253. Springer, Cham (2020). https://doi.org/10.1007/978-3-030-39074-7_26

18. Zhang, J., Xie, Y., Zhang, P., Chen, H., Xia, Y., Shen, C.: Light-weight hybrid convolutional network for liver tumor segmentation. In: IJCAI, pp. 4271–4277 (2019)

19. Zheng, R., Zhao, X., Zhao, X., Wang, H.: Deep learning based multi-modal cardiac MR image segmentation. In: Pop, M., et al. (eds.) STACOM 2019. LNCS, vol. 12009, pp. 263–270. Springer, Cham (2020). https://doi.org/10.1007/978-3-030-39074-7_28

20. Zhuang, X.: Multivariate mixture model for cardiac segmentation from multi-sequence MRI. In: Ourselin, S., Joskowicz, L., Sabuncu, M.R., Unal, G., Wells, W. (eds.) MICCAI 2016. LNCS, vol. 9901, pp. 581–588. Springer, Cham (2016). https://doi.org/10.1007/978-3-319-46723-8_67

21. Zhuang, X.: Multivariate mixture model for myocardial segmentation combining multi-source images. IEEE Trans. Pattern Anal. Mach. Intell. **41**(12), 2933–2946 (2019)

Two-Stage Method for Segmentation of the Myocardial Scars and Edema on Multi-sequence Cardiac Magnetic Resonance

Yanfei Liu, Maodan Zhang, Qi Zhan, Dongdong Gu, and Guocai Liu[✉]

Hunan University, Changsha, China
lgc630819@hnu.edu.cn

Abstract. Segmenting cardiac scars and edema from cardiac magnetic resonance (CMR) are essential for the early diagnosis and accurate prognostic assessment of ischemic heart disease. The pathological myocardium presents distinctive brightness in the late gadolinium enhancement (LGE) images, the T2-weighted CMR shows the acute injury and ischemic regions, and the balanced-Steady State Free Precession (bSSFP) can clearly reveal the boundaries of the myocardium. Given this fact, we proposed a novel fully-automatic two-stage method to extract different features of each modality as well as segment myocardium edema and scars. In the first stage, a U-net was trained on bSSFP images with full annotation of myocardium, which can locate the coarse position of the myocardium and obtain the mask of the myocardium as a constraint on the next stage. In the second stage, with the T2 images, LGE images and predicted myocardium masks concatenated as inputs, an M-shaped network based on attention mechanism was trained to segment the myocardial edema and scars accurately. In conclusion, the accuracy of the segmentation was improved by adopting prior constraints and attention mechanism, which achieved an average Dice score of 0.570 and 0.634 for the myocardial scars and myocardial scars+edema respectively on the test set of MyoPS 2020.

Keywords: Segmentation · Multi-sequence CMR · Deep learning · Attention mechanism

1 Introduction

Cardiac magnetic resonance (CMR) images can provide anatomical and functional information of the heart, which is crucial to clinical diagnosis and the treatment of myocardial infarction. To this end, the segmentation of the myocardial pathology is a critical step for the analysis of myocardial infarction. Different

---

The original version of this chapter was revised: the acknowledgement and Dongdong Gu's affiliations were updated. The correction to this chapter is available at https://doi.org/10.1007/978-3-030-65651-5_17

© Springer Nature Switzerland AG 2020, corrected publication 2021
X. Zhuang and L. Li (Eds.): MyoPS 2020, LNCS 12554, pp. 26–36, 2020.
https://doi.org/10.1007/978-3-030-65651-5_3

CMR sequences have different focuses: balanced-Steady State Free Precession (bSSFP) images can present the complete myocardial boundary, T2-weighted images show myocardial edema clearly, and late gadolinium enhancement (LGE) images highlight the myocardial scars. By combining multiple sequences, rich and reliable information about the pathology and morphology of the myocardium can be obtained. In the absence of a unified automatic segmentation standard in clinical practice, the segmentation process is usually done manually, which is time-consuming and depends on inter- and intra-observer variations.

In the existing literatures, traditional machine learning technology showed good performance in cardiac image segmentation [4,7], but it required manual feature selection or prior knowledge to achieve satisfactory accuracy. In contrast, deep learning algorithms can automatically extract features. As the number of public datasets has increased in recent years, many deep learning-based segmentation algorithms have been developed for CMR. As a type of convolutional neural network (CNN) without any fully-connected layers, the fully convolutional neural network (FCN) [5] and U-Net with skip-connections are adopted frequently in many other methods [6,10]. Many of the works have been focused on the segmentation of the cardiac chambers, with relatively fewer studies on segmenting abnormal myocardial tissue regions, such as left ventricle (LV) scars and edema. In [1,11], the authors accurately quantify the amount of scars in patients suffering from ischemic and hypertrophic cardiomyopathy respectively. However, these works do not focus on the segmentation of myocardial scars and edema in multi-sequence CMR.

In this paper, we proposed a novel fully-automatic two-stage method to segment myocardial scars and edema in multi-sequence CMR. Our method mainly consists of two neural networks:

- A segmentation network for bFFSP images: A U-net is used to roughly locate and segment the entire myocardium. The mask of the myocardium obtained at this stage is applied as part of the input of the second stage to constrain the location of myocardial scars and edema.
- A segmentation network for multi-sequence images: An M-shaped network utilizes the constraints of myocardial shape and the attention mechanism is applied to segment both myocardial scars and edema.

2   Method

2.1   Dataset Description

The dataset is provided by myocardial pathology segmentation combining multi-sequence CMR (Myops20) [12,13] including 45 cases of three sequence CMR: bSSFP, T2 and LGE, of which 25 cases are with annotations. For the original CMR sequence of each patient, the bSSFP images consist of 8–12 slices, with in-plane resolution of 1.25×1.25 mm and slice thickness of 8 to 13 mm. The T2 images consist of 3–7 slices, with in-plane resolution of 1.35×1.35 mm and slice thickness of 12 to 20 mm. The LGE images have 10 18 slices with in-plane resolution of 0.75×0.75 mm and slice thickness of 5 mm. The images are aligned into a common space and resampled into the same spatial resolution. The provided gold standard labels are non-overlapping, and they include: left ventricular blood pool (labelled 500), right ventricular blood pool (labelled 600), LV normal myocardium (labelled 200), LV myocardial edema (labelled 1220), and LV myocardial scars (labelled 2221). In addition, this segmentation task can be more difficult than others with selected dataset, because the dataset is directly collected from the clinic without any selection.

2.2 Image Segmentation

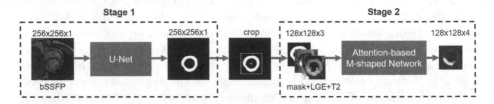

Fig. 1. Overview of the two-stage segmentation method. In the first stage, the method uses bSSFP images and U-net to obtain the prior constraint. The ouput is cropped to the size of 128×128 based on it. The input in the second stage is the prior constraint concatenated with LGE and T2 images. The attention-based M-shaped network is used to segment the myocardial scars and edema.

In order to facilitate training, we crop images into the size of 256×256 uniformly as the sizes for each case are different, and to ensure the heart is roughly at the center of the image. Due to the low proportion of the myocardial edema and myocardial scars region in images, as the labels provided are non-overlapping, the labels of LV normal myocardium (200), LV myocardial edema (1220) and LV myocardial scars (2221) are combined as the approximate labels of the entire LV myocardial tissue in the first stage, see Fig. 2 (Fig. 1).

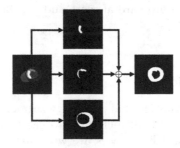

Fig. 2. Combination of the labels in stage 1. The labels of LV normal myocardium (200), LV myocardial edema (1220) and LV myocardial scars (2221) are set to 1 and other pixels are set to 0.

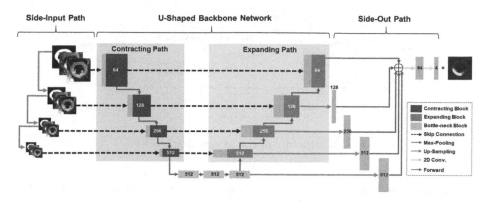

Fig. 3. Overview of the M-shaped segmentation network. The number in the boxes corresponds to the number of channels of the block.

In the first stage, a U-net is trained with the bSSFP images and the approximate myocardial labels to obtain the position of the myocardium roughly. By combining three classes into one, the errors from the class-imbalance problem can be alleviated. Furthermore, the predicted label of the myocardium is taken as the prior constraint, which becomes part of the input in the second stage.

According to the center of the predicted myocardial mask, the images are re-cropped to the size of 128×128, because the myocardium is a circular tissue. Even with the same sequence, the data range of the cases for different patients varies greatly, so histogram equalization and random gamma technique are applied for the cropped images to balance the data distribution after setting the window level and the window width uniformly. Furthermore, the common data augmentation strategies including random rotation, random crop and random scaling are utilized for training data.

In the second stage, the input data is stacked by T2 images, LGE images and the myocardium labels obtained in the first stage. Inspired by [2], our network is expanded on the basis of U-net, consisting of a side-input path, adandan U-shape backbone network, and a side-output layer. The U-shaped convolutional network is employed as the main body structure consisting of contracting blocks, expanding blocks and bottleneck blocks, shown in Fig. 3. In each block, the convolution module consists of a convolutional layer with filter size of 3×3 and L1 regularization, a Rectifier Linear Unit (ReLU) layer and a spatial dropout layer to prevent the overfitting.

In the contracting path, the contracting block is a structure that combines the channel attention module from convolutional block attention module (CBAM) [9] and the residual connection [3]. Figure 4. displays the structure of the contracting block. In the channel attention module, two different spatial context descriptors are obtained to compress the feature map in spatial dimensions by using maximum pooling and average pooling. Composed of multi-layer perceptron (MLP), the shared network is used to calculate the two different spatial context descriptors to obtain the channel attention map. The channel attention

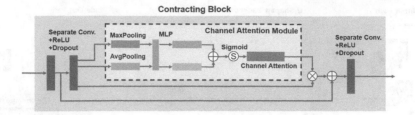

Fig. 4. Structure of the contracting block.

map is multiplied by the input of the channel attention module, added to the output of the first convolutional layer of the contracting block to form a residual structure, and followed by a convolutional layer. The convolution layer in the contracting block is replaced by the separable convolution layer to better extract the features. In addition, in the final of the contracting block, the convolution layer with the stride of 2 is utilized to down-sample the feature maps.

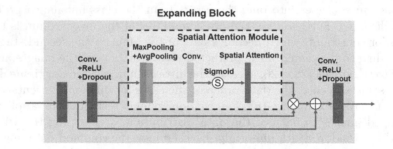

Fig. 5. Structure of the expanding block.

Figure 5 illustrates the structure of the expanding block which is similar to the contracting block, except the spatial attention module from CBAM and the up-sampling layer for output. Different from channel attention, spatial attention mainly focuses on location information. In order to calculate spatial attention, maximum pooling and average pooling are adopted in the dimension of the channel to obtain two different feature maps, after which they are concatenated and convolved by a convolutional layer.

The bottle-neck block takes advantages of the full CBAM block and residual connections, which can learn the hierarchical representation and extract informative features (see Fig. 6.).

The architecture includes skip connections between all contracting blocks and expanding blocks at the same spatial resolution. Therefore, the high-level global information and low-level details can be combined and taken into account. The side-input path is an image pyramid by maximum pooling layers with the stride size of 2 to obtain multi-scale receptive fields. Each layer of side-input path is

Bottle-neck Block

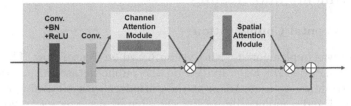

Fig. 6. Structure of the bottle-neck block.

concatenated with the contracting block which has the same spatial resolution. The side-output path consists of feature maps with the same resolution, which is up-sampled from different blocks in the expanding path. In this way, while ensuring the width of the network, the gradient disappearance can be alleviated. On the other hand, the multi-scale output images can be supervised to obtain a better segmentation result. The network outputs four classes of segmentation results after the Softmax activation layer: the myocardial scars, the myocardial edema, the normal myocardium and background.

To train the network, we employ a composite loss function L_{seg} that consists of two loss terms: $L_{seg} = L_{FDL} + \lambda L_{mse}$. As the extreme class imbalance exists among different labels, we use the focal dice loss L_{FDL} as the first term in [8]:

$$L_{FDL} = \sum_t w_t(1 - Dice_t^{1/\beta})$$

where w indicates the weight for each class t, and the factor $1/\beta$ represents the power of $Dice_t$ for each class. $Dice$ coefficient is a measure of the similarity between the prediction and ground truth, which is defined as follows:

$$Dice_t = \frac{2|P_t \cap G_t|}{P_t + G_t}$$

where P and G represent the predicted labels and ground truth respectively. The classes that are difficult to segment can get higher weights in the segmentation process, so the network can focus on the learning of the more difficult classes. We set $w = \{1, 1, 1, 0.5\}$ for the four classes respectively and $\beta = 2$. The second term L_{mse} is adopted to optimize the edge of the $H \times W$ predicted segmentation results and labels:

$$L_{mse} = \frac{1}{H \times W} \sum_{i=1}^{H} \sum_{j=1}^{W} (P_t(i,j) - G_t(i,j))^2$$

We set $\lambda = 100$ to balance the contribution of the two losses according our experience.

3   Experiments and Results

3.1   Experimental Configuration

Our network was implemented with Python based on TensorFlow. The training set and the validation set were divided on the public dataset at a ratio of 4:1. During the training, we used Adam as the optimizer with a learning rate starting from 3e−4 and a momentum of 0.5. The batch size was set to 20. A fixed threshold 0.5 was employed to get a binary mask from the probability map. The U-net in the first stage was trained for 600 epochs and the M-shaped network in the second stage was trained for 800 epochs on an NVIDIA TITAN V GPU.

3.2   Performance Evaluation and Analysis

Network Structure. In order to define the best trained framework for the segmentation of myocardial scars and edema, we compared various structures in the second stage: (1) a single U-net, (2) a U-shape backbone network (our M-shaped network without the side input/output path), (3) an M-shaped network (Our M-shaped network without attention mechanism) and (4) an M-shaped network + CBAM (our method). In terms of Table 1, it is obvious that in the validation set of 5 patients, our method leads to the best segmentation, achieving the dice score of 0.638 ± 0.097 for myocardial scars and 0.561 ± 0.103 for myocardial scars+edema.

Table 1. Average and standard deviation for the Dice score of different structures on validation set of 5 patients.

	scars		scars+edema	
	avg.	std.	avg.	std.
U-net	0.560	0.132	0.413	0.156
M-shaped network	0.571	0.142	0.512	**0.068**
U-shaped network+CBAM	0.607	0.133	0.549	0.079
M-shaped network+CBAM	**0.638**	**0.097**	**0.561**	0.103

Figure 7 presents the results of different structures. By the comparison of (1) and (2), we know that the residual connection can improve the accuracy of segmentation to a certain extent. The comparison of (2) and (4) proves that CBAM block can improve the Dice score for the myocardial scars by ∼12%. The comparison of (3) and (4) shows that the network performs better in details with the side-input/output path. Finally, the comparison between (1) and (4) reveals that, combined with the attention mechanism and the side-input/output path, the residual structure plays an important role in the improvement of the segmentation performance.

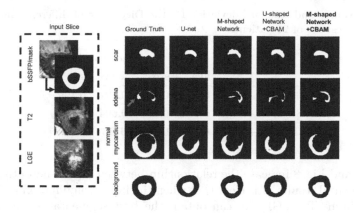

Fig. 7. The segmentation results of an example by using different structures. The leftmost column are the input slices. Each column on the right has four classes segmentation results, corresponding to different structures.

Prior Constraint. In order to prove the validity of the prior constraints, we designed experiments for the training process and the inference process respectively.

In the training process, we compared different input combinations: (1) T2+LGE, (2) bSSFP+T2+LGE, and (3) the mask of myocardium+T2+LGE. The different combinations were trained by the method in the second stage.

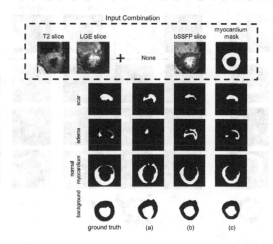

Fig. 8. Examples of segmentation results by adopting different input combinations in the training process. The top row exhibits three different combinations: (1) T2+LGE, (2) bSSFP+T2+LGE, and (3) the mask of myocardium+T2+LGE. Each column below is four classes segmentation results corresponding to different combinations.

Table 2. Average and standard deviation for the Dice score of difference input combinations on test set of 20 patients.

	scars		scars+edema	
	avg.	std.	avg.	std.
T2+LGE	0.516	**0.246**	0.527	0.199
bSSFP+T2+LGE	0.553	0.277	0.604	0.165
mask+T2+LGE	**0.570**	0.283	**0.634**	**0.164**

Table 2 and Fig. 8 indicate the relationship between input and output. From the experimental results, it is clear that when the input combination is the mask of myocardium+T2+LGE, we can obtain the best segmentation results. The boundary information of the myocardial region provided by the bSSFP images is significant, but the role is limited without the extraction of information from bSSFP images.

In the inference process, the trained network in the second stage was used to evaluate the effectiveness of the constraint by changing the channel of the constraint on the basis of the unchanged T2 and LGE channels: (1) all black (all pixels are 0), (2) all white (all pixels are 1), (3) an example of a bad myocardium mask, (4) another example of a bad myocardium mask, and (5) an example of a good myocardium mask (Fig. 9).

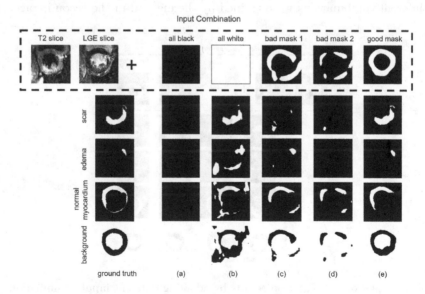

Fig. 9. Examples of segmentation results using different input combinations in the inference process. The top row displays five different combinations: (1) all black+T2+LGE, (2) all white+T2+LGE, (3) a bad mask+T2+LGE, (4) another bad mask+T2+LGE, and (5) a good mask+T2+LGE. Each column below has four classes segmentation results corresponding to different combinations.

The experimental results illustrate that the constraints have a great influence on the output of the network. The segmentation results in each part of the myocardium are closely related to the input constraint information. When the pixels of the constraint channel are all 0, neither myocardial scar nor edema has an output, indicating that they are restrained. When the pixels of the constraint channel are all 1, the output of the network is messy, which proves that the prediction of myocardial scars and edema is constrained by shape at the same time. The examples of bad masks can also support this point.

4   Conclusion

This paper proposed a two-stage segmentation method for the myocardial scars and edema on multi-sequence CMR that includes two networks. In the first stage, a U-net is applied to extract the entire myocardial part to obtain the prior constraint. In the second stage, an M-shaped segmentation network based on the attention mechanism and residual connection is adopted, which can improve segmentation accuracy of myocardial scars and edema. The method utilizes different characteristics of each sequence of CMR through different stages, and integrates the prior constraint of myocardium into the training process to achieve the better segmentation results. The method achieves a Dice score of 0.570 ± 0.283 for myocardial scars and 0.634 ± 0.164 for myocardial scars+edema on the test set of MyoPS20.

Acknowledgement. This work was supported by the National Natural Science Foundation of China (61671204).

References

1. Fahmy, A.S., et al.: Three-dimensional deep convolutional neural networks for automated myocardial scar quantification in hypertrophic cardiomyopathy: a multicenter multivendor study. Radiology **294**(1), 52–60 (2020)
2. Fu, H., Cheng, J., Xu, Y., Wong, D.W.K., Liu, J., Cao, X.: Joint optic disc and cup segmentation based on multi-label deep network and polar transformation. IEEE Trans. Med. Imaging **37**(7), 1597–1605 (2018)
3. He, K., Zhang, X., Ren, S., Sun, J.: Deep residual learning for image recognition. In: Proceedings of the IEEE Conference on Computer Vision and Pattern Recognition, pp. 770–778 (2016)
4. Lin, A., Kolossváry, M., Išgum, I., Maurovich-Horvat, P., Slomka, P.J., Dey, D.: Artificial intelligence: improving the efficiency of cardiovascular imaging. Expert. Rev. Med. Devices **17**(6), 565–577 (2020). https://doi.org/10.1080/17434440.2020.1777855
5. Long, J., Shelhamer, E., Darrell, T.: Fully convolutional networks for semantic segmentation. In: Proceedings of the IEEE Conference on Computer Vision and Pattern Recognition, pp. 3431–3440 (2015)
6. Singh, G., et al.: Deep learning based automatic segmentation of cardiac computed tomography. J. Am. Coll. Cardiol. **73**(9), 1643 (2019)

7. Tavakoli, V., Amini, A.A.: A survey of shaped-based registration and segmentation techniques for cardiac images. Comput. Vis. Image Underst. **117**(9), 966–989 (2013)
8. Wang, P., Chung, A.C.S.: Focal dice loss and image dilation for brain tumor segmentation. In: Stoyanov, D., et al. (eds.) DLMIA/ML-CDS -2018. LNCS, vol. 11045, pp. 119–127. Springer, Cham (2018). https://doi.org/10.1007/978-3-030-00889-5_14
9. Woo, S., Park, J., Lee, J.-Y., Kweon, I.S.: CBAM: convolutional block attention module. In: Ferrari, V., Hebert, M., Sminchisescu, C., Weiss, Y. (eds.) ECCV 2018. LNCS, vol. 11211, pp. 3–19. Springer, Cham (2018). https://doi.org/10.1007/978-3-030-01234-2_1
10. Xu, Z., Wu, Z., Feng, J.: CFUN: combining faster R-CNN and U-net network for efficient whole heart segmentation. arXiv preprint arXiv:1812.04914 (2018)
11. Zabihollahy, F., White, J.A., Ukwatta, E.: Fully automated segmentation of left ventricular myocardium from 3D late gadolinium enhancement magnetic resonance images using a U-net convolutional neural network-based model. In: Medical Imaging 2019: Computer-Aided Diagnosis, vol. 10950, p. 109503C. International Society for Optics and Photonics (2019)
12. Zhuang, X.: Multivariate mixture model for cardiac segmentation from multi-sequence MRI. In: Ourselin, S., Joskowicz, L., Sabuncu, M.R., Unal, G., Wells, W. (eds.) MICCAI 2016. LNCS, vol. 9901, pp. 581–588. Springer, Cham (2016). https://doi.org/10.1007/978-3-319-46723-8_67
13. Zhuang, X.: Multivariate mixture model for myocardial segmentation combining multi-source images. IEEE Trans. Pattern Anal. Mach. Intell. **41**(12), 2933–2946 (2019)

Multi-modality Pathology Segmentation Framework: Application to Cardiac Magnetic Resonance Images

Zhen Zhang[1], Chenyu Liu[1], Wangbin Ding[1], Sihan Wang[2], Chenhao Pei[1], Mingjing Yang[1(✉)], and Liqin Huang[1]

[1] College of Physics and Information Engineering, Fuzhou University, Fuzhou, China
`yangmj5@fzu.edu.cn`
[2] School of Basic Medical Science, Fudan University, Shanghai, China

Abstract. Multi-sequence of cardiac magnetic resonance (CMR) images can provide complementary information for myocardial pathology (scar and edema). However, it is still challenging to fuse these underlying information for pathology segmentation effectively. This work presents an automatic cascade pathology segmentation framework based on multi-modality CMR images. It mainly consists of two neural networks: an anatomical structure segmentation network (ASSN) and a pathological region segmentation network (PRSN). Specifically, the ASSN aims to segment the anatomical structure where the pathology may exist, and it can provide a spatial prior for the pathological region segmentation. In addition, we integrate a denoising auto-encoder (DAE) into the ASSN to generate segmentation results with plausible shapes. The PRSN is designed to segment pathological region based on the result of ASSN, in which a fusion block based on channel attention is proposed to better aggregate multi-modality information from multi-modality CMR images. Experiments from the MyoPS2020 challenge dataset show that our framework can achieve promising performance for myocardial scar and edema segmentation.

Keywords: Myocardial pathology · Multi-sequence CMR · Segmentation

1 Introduction

Myocardial infarction (MI) is one of the most dangerous cardiovascular diseases in worldwide. The severity of MI depends on the assessment of the myocardial scar and edema [3]. Accurate delineation of these pathological regions from cardiac magnetic resonance (CMR) can provide important advancements for the prediction and management of MI patients [5]. Since manual delineation is generally time-consuming, tedious and subject to inter-observer variations, the automatic segmentation approach has gradually attracted more attention of research.

© Springer Nature Switzerland AG 2020
X. Zhuang and L. Li (Eds.): MyoPS 2020, LNCS 12554, pp. 37–48, 2020.
https://doi.org/10.1007/978-3-030-65651-5_4

Conventional myocardial pathology segmentation methods are mainly based on intensity thresholding, such as the signal threshold to reference mean (STRM) [6], region growing (RG) [1] and full-width at half-maximum (FWHM) [2]. However, the thresholding methods could be easily affected by the image noise, and have poor agreement with expert delineations [9,16]. Recently, learning-based methods have achieved promising performance in different pathology segmentation tasks, such as brain tumor [10] and liver lesion [15]. For pathology segmentation on left atrium (LA) myocardium, Yang et al. presented a super-pixel scar segmentation method using support vector machine (SVM) [11]. Li et al. proposed a fully automated scar segmentation method based on the graph-cuts framework, where the potentials of the graph are estimated via deep neural network (DNN) [8]. Futhermore, Li et al. designed a multi-task learning network to joint perform LA segmentation and LA scars quantification, in which the LA boundary is extracted as spatial attention for the scars [7]. For pathology segmentation on left ventricular (LV) myocardium, Zabihollahy et al. proposed a CNN-based method to segment scar from late gadolinium enhancement (LGE) MRIs [14]. However, their method relies on the manual delineation of the LV myocardium region. To achieve a fully automatic scar segmentation method, they further developed a multi-planar network to segment LV myocardium [13].

At present, most DNN-based myocardial pathology segmentation methods are focus on mono-modality CMR, such as LGE. But multi-modality CMR can provide different enhanced-information of the whole heart. For instance, the balanced-Steady State Free Precession (bSSFP) cine sequence can present a clear myocardial boundary, while the LGE and T2-weighted CMR can highlight the scar and edema regions, respectively [19,20]. Being aware that the complementary information is helpful for myocardial pathology segmentation. We design a cascade multi-modality pathology segmentation framework. Figure 1 shows the overview of the framework. The framework decomposes the pathology segmentation task into two sub-stages, i.e. the anatomical structure segmentation stage and the pathological region segmentation stage. The main contributions of our work are:

(1) we propose a fully automatic pathology segmentation framework, and validate it using the MyoPS2020 challenge dataset [1].
(2) we present an anatomical structure segmentation network, where a denoising auto-encoder (DAE) is adopted to reconstruct the segmentation results with realistic shapes.
(3) we propose a pathological region segmentation network, in which a channel attention based fusion block is designed to adaptively fuse complementary information of multi-modality CMR images for pathology segmentation.

---

[1] http://www.sdspeople.fudan.edu.cn/zhuangxiahai/0/MyoPS20/.

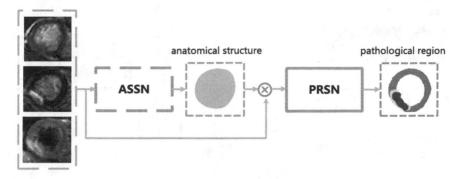

Fig. 1. The architecture of the multi-modality pathology segmentation framework. Given multi-modality CMR (bSSFP, T2, DE) images, the ASSN first obtains a candidate anatomical structure, where the pathology may exist. Then, the PRSN predicts the final scar and edema regions within the candidate structure.

2   Method

2.1   Anatomical Structure Segmentation Network (ASSN)

The ASSN is designed to obtain a candidate anatomical structure from CMR images. In the myocardial pathology segmentation task, we designate the candidate structure as the LV epicardial region, where the scar and edema may exist. Figure 2 shows the architecture of the ASSN. It mainly includes three individual encoders and one shared decoder. Each encoder can obtain underlying anatomical feature from CMR, while the decoder can fuse the obtained features, and predicts a pixel-level LV mask.

Given a multi-modality CMR images $I = (I_{bSSFP}, I_{LGE}, I_{T2})$, the ASSN aims to learn a mapping f_θ from I to a binary mask. Therefore, the network can be trained under supervised manner, and the loss function is

$$\mathcal{L}oss_{seg} = \mathcal{D}ice(f_\theta(I), L_{lv}) \tag{1}$$

where the L_{lv} is the golden standard of the LV, $\mathcal{D}ice(A, B)$ refers to the Dice score of A and B. Thus, the candidate anatomical structure $C = (C_{bSSPF}, C_{LGE}, C_{T2})$ can be extracted as

$$C_{bSSFP} = I_{bSSFP} \otimes f_\theta(I), \tag{2}$$

$$C_{LGE} = I_{LEG} \otimes f_\theta(I), \tag{3}$$

$$C_{T2} = I_{T2} \otimes f_\theta(I), \tag{4}$$

where $\otimes$ is element-wise multiplication.

Generally, the ASSN performs pixel wise classification based on processing the intensity value of I. However, the pathology usually leads to abnormal intensity distribution in CMR images. For instance, LGE visualizes the scars as

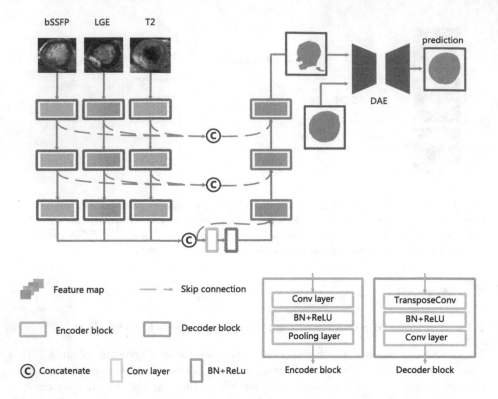

Fig. 2. The architecture of the ASSN. The auxiliary DAE is adopted to suppress the influence of the pathology region and generate results with plausible shape.

brighter texture, in contrast to the dark healthy myocardium [21]. Therefore, the segmentation results could be easily affected. To tackle this, we adopt a DAE to refine the segmentation results with realistic shapes [12].

A DAE usually follows an encoder-decoder (E-D) architecture. Let the $\ddot{L}$ denotes the noisy version of the L_{gd}, the DAE aims to map the $\ddot{L}$ to a lower-dimension representation h, from which the L_{gd} can be reconstructed. It can be trained to minimize the reconstruction error of the input

$$Loss_{DAE} = \|D(E(\ddot{L})) - L_{gd}\|_2, \tag{5}$$

where $E(\ddot{L})$ is a compact representation of $\ddot{L}$, $D(E(\ddot{L}))$ is a reconstruction of L_{gd}. Regarding the original segmentation result $f_\theta(I)$ as a noisy version of the golden standard label, we integrate the DAE into the ASSN to reconstruct the original result into a plausible one. So that, the final loss function of the ASSN is defined as

$$Loss_{ASSN} = Loss_{seg} + \beta Dice(D(E(f_\theta(I))), L_{lv}), \tag{6}$$

where β is the balance coefficient between the Dice loss and reconstruction loss.

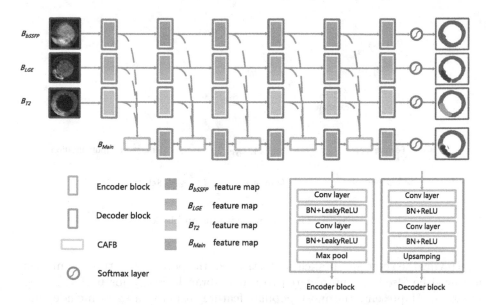

Fig. 3. The architecture of PRSN. It contains four sub-branches B_{bSSFP}, B_{LEG}, B_{T2}, B_{Main}. We set up to segment the myocardium in B_{bSSFP}; the scars and normal myocardium in B_{LEG}; the edema and normal myocardium in B_{T2}; the scars, edema and normal myocardium in B_{Main}.

2.2 Pathological Region Segmentation Network (PRSN)

In the pathological region segmentation network (PRSN), complementary information from $C = (C_{bSSPF}, C_{LGE}, C_{T2})$ are expected to be fused and boost the pathology segmentation performance. Figure 3 shows the architecture of PRSN. We construct three DNN branches $(B_{bSSPF}, B_{LGE}, B_{T2})$ to capture multi-modality information from each candidate region. Specifically, the B_{LEG} and B_{T2} mainly aim to acquire pathology (scar and edema) features from C_{LGE} and C_{T2}. Meanwhile, the B_{bSSFP} is intent to obtain myocardium features from C_{bSSPF}. Due to most of the scar and edema are scatted in the myocardium, the myocardium features can provide spatial prior information for the pathology regions. Therefore, the cost functions of three DNN branches are

$$\mathcal{L}oss_{bSSFP} = \mathcal{D}ice(L_{bSSFP}, \hat{L}_{bSSFP}), \tag{7}$$

$$\mathcal{L}oss_{LGE} = \mathcal{D}ice(L_{LGE}, \hat{L}_{LGE}), \tag{8}$$

$$\mathcal{L}oss_{T2} = \mathcal{D}ice(L_{T2}, \hat{L}_{T2}), \tag{9}$$

where L_{bSSFP} (L_{LGE}, L_{T2}) and $\hat{L}_{bSSFP}$ ($\hat{L}_{LGE}$, $\hat{L}_{T2}$) are corresponding gold standard and predicted label of B_{bSSPF} (B_{LEG}, B_{T2}) branch, respectively.

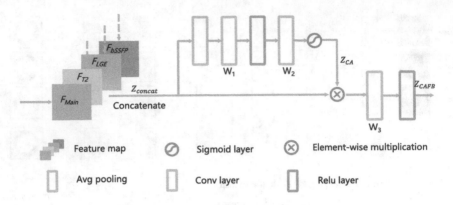

Fig. 4. The channel-wise fusion block

Having the three sub-branches constructed, the potential features from them are need be fused and propagated to a main-branch B_{Main} for pathology segmenting. At present, the most popular feature fusion strategies include summation, product and maximization. However, they still suffer from the lack of robustness in different tasks [17]. As shown in the Fig. 4, we propose a multi-modality channel-attention fusion block (CAFB) for adaptively weighted feature fusion of different modalities in B_{Main}.

Suppose we have three feature maps $(F_{bSSFP}, F_{LGE}, F_{T2})$ from sub-branches and one previous output F_{Main} of the main-branch, the CAFB first merges these feature maps to obtain an concatenated feature Z_{concat}. Since Z_{concat} aggregates all feature maps from $(B_{bSSFP}, B_{LEG}, B_{T2})$, it easily suffers from the information redundancy. Due to the channel-attention (CA) [4] can emphasize informative features and suppress less useful ones, the block adopts it to performs channel-wise feature re-calibration on $Z_{concate}$. So That, the output of CAFB (Z_{CAFB}) is

$$Z_{CA} = \delta(W_2\sigma(W_1 Avg(Z_{concat})), \tag{10}$$

$$Z_{CAFB} = \sigma(W_3(Z_{CA} \otimes Z_{concat})), \tag{11}$$

where σ, δ and Avg refer to Relu, Sigmoid and average pooling function, respectively; and W_1, W_2 and W_3 are parameters of different convolution layers. Here, Z_{CA} is the channel-wise attention weight, with which the original Z_{concat} can be re-calibrated and achieve better representation of multi-modality information.

Furthermore, we apply the CAFB in different hierarchies of the B_{Main} (see B_{Main} in Fig. 3). Thus, the B_{Main} can capture multi-scale multi-modality features for pathology segmentation. The training loss of the B_{Main} can be defined as

$$\mathcal{L}oss_{Main} = \mathcal{D}ice(L_{Main}, \hat{L}_{Main}), \tag{12}$$

where L_{Main} and $\hat{L}_{Main}$ are the gold standard and predicted label of B_{Main}, respectively. Note that the B_{Main} jointly performs scar, edema and normal myocardium segmentation. Finally, the overall loss function of the PRSN is

$$Loss_{PRSN} = Loss_{Main} + \lambda_{bSSFP} Loss_{bSSFP} + \lambda_{LGE} Loss_{LEG} + \lambda_{T2} Loss_{T2}, \quad (13)$$

where λ_{bSSFP}, λ_{LGE} and λ_{T2} are hyper-parameters.

3 Experiment

3.1 Dataset

The framework was trained and evaluated in the MyoPS2020 challenge data set which contains 25 labelled and 20 unlabelled multi-sequence CMR (bSSFP, LGE, T2) images. All published data has been aligned in a common space and resampled with the same spatial resolution. In our experiments, we randomly selected 20 labelled images for network training, while leaving the rest of 5 labelled samples for validation, and the final performance of the framework was evaluated on the 20 unlabelled images.

3.2 Implementations

We trained our models by extracting 2D slices from multi-sequence CMR images. Each slice was cropped and resized to 128×128 pixels which are roughly centering at the heart region. All of the models (DAE, ASSN and PRSN) were implemented in Python and optimized by using the Adam algorithm.

For the DAE: In each training iteration, we generated $\ddot{L}$ by randomly adding noise to a gold standard label L. Having a pair of $(L, \ddot{L})$ prepared, the DAE can be trained via minimizing the reconstruction loss $Loss_{DAE}$ (see Eq. 5).

For the ASSN: The pre-trained DAE was adopted to perform shape reconstruction. In each training iteration, the sample $I = (I_{bSSFP}, I_{LGE}, I_{T2}, L)$ was feed into the network. By setting the β to 0.2 in $\mathcal{L}_{ASSN}$ (see Eq. 6), the trainable loss can be calculated and back-propagated to optimize the parameters of ASSN.

For the PRSN: We first extracted $C = (C_{bSSPF}, C_{LGE}, C_{T2})$ and their corresponding label $(L_{bSSPF}, L_{LGE}, L_{T2})$ from the training data. Then, the hyperparameter λ_{bSSFP}, λ_{LGE} and λ_{T2} were set to 0.3, 0.5 and 0.5, respectively (see Eq. 13). Finally, the network can be trained by minimizing the $Loss_{PRSN}$.

3.3 Results

ASSN: To evaluate the performance of the ASSN, the Dice score and Hausdorff distance between the predicted label and gold standard label were calculated. Table 1 shows the performance of three different methods:

- Unet-bSSFP. The Unet which is trained by using bSSFP images. We implemented this method because the bSSFP can provide a relatively clear boundary of the LV.

- ASSN-WO-DAE: Our ASSN network but without DAE.
- ASSN: Our proposed anatomical structure segmentation network.

Compared to Unet-bSSFP, the methods (ASSN-WO-DAE and ASSN) using multi-modality CMRs can achieve better performance in both terms of Dice score and Hausdorff distance. Additionally, although the ASSN-WO-DAE obtained comparable result to ASSN in term of the Dice score, the ASSN still achieved almost 5 mm improvement in the Hausdorff distance. Moreover, Fig. 5 presents a series of visual results. One can see the results of Unet-bSSFP and ASSN-WO-DAE were easily affected by the quality of CMRs, while our ASSN can generate results with plausible shape. This demonstrates the benefit of integrating DAE to the segmentation network.

Table 1. Dice score and Hausdorff of the proposed method and other baseline methods on the validation set.

Method	Dice (%)	Hausdorff (mm)
Unet-bSSFP	93.77 ± 2.55	10.72 ± 9.02
ASSN-WO-DAE	96.21 ± 3.28	8.25 ± 9.06
ASSN	**96.99 ± 1.49**	**3.04 ± 0.76**

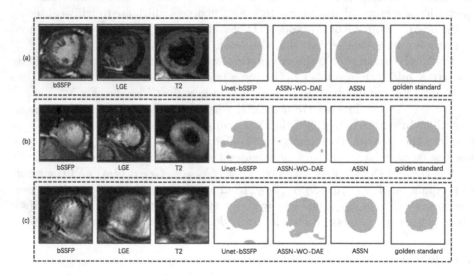

Fig. 5. Comparison of the proposed method and other baseline methods. Image of (a) is a normal case, where all three methods can achieve a reasonable segmentation result. However, the image of (b) and (c) are challenging cases, where both Unet-bSSFP and ASSN-WO-DAE failed to generate realistic results but the proposed ASSN method showed good robustness.

PRSN: The performance of the PRSN is evaluated by the Dice score of scar and scar + edema region. Table 2 shows seven different segmentation methods based on our extracted LV.

- Unet-scar: Unet which is trained on C_{LGE} datas for scar segmentation.
- Unet-edema: Unet which is trained on C_{T2} datas for edema segmentation.
- PRSN-B_{T2}: The B_{T2} branch of PRSN.
- PRSN-B_{LGE}: The B_{LGE} branch of PRSN.
- Fusion-Unet: Unet which is implemented by using input-level fusion strategy [18].
- MFB-PRSN: PRSN which is implemented by using MFB (summation-product-maximization) fusion strategy in B_{Main} [17].
- PRSN: Our proposed pathological region segmentation network.

Among these methods, the Unet-scar, Unet-edema, PRSN-B_{T2} and PRSN-B_{LGE} can be considered as the mono-modality methods, while the MFB-PRSN, Fusion-Unet and PRSN are multi-modality methods. Overall, the multi-modality methods achieved better results than the mono-modality methods in scar segmentation. This reveals the advantage of using multi-modality images for pathology segmentation. Meanwhile, compared to MFB-PRSN which uses the summation-product-maximization fusion strategy for feature fusion, our PRSN achieved almost 4% and 3% improvement in scar and scar + edema region, respectively. This indicates the advantage of our proposed CAFB. In addition, Fig. 6 demonstrates visual results of different methods.

Table 2. Dice scores of the proposed method and other baseline methods on the testing set. N/A indicates the segmentation result was not provided

Method	scar (%)	edema (%)
Unet-edema	N/A	61.42 ± 11.86
Unet-scar	56.38 ± 23.36	N/A
PRSN-B_{T2}	N/A	64.37 ± 11.25
PRSN-B_{LGE}	55.65 ± 24.97	N/A
	scar (%)	scar + edema (%)
Fusion-Unet	57.50 ± 23.09	66.61 ± 12.79
MFB-PRSN	59.56 ± 25.18	68.59 ± 12.33
PRSN	**64.09 ± 25.96**	**70.24 ± 12.98**

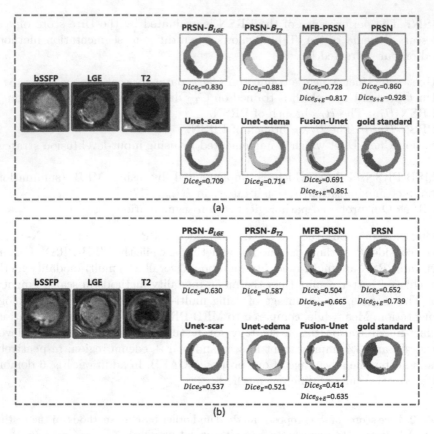

Fig. 6. Visualization of the pathology segmentation results. The $Dice_S$, $Dice_E$ and $Dice_{S+E}$ refer to the Dice score of predicted scar (blue), edema (green) and scar + edema region, respectively. Image of (a) is an easy case, while the image of (b) is a more challenging one. (The reader is referred to the colourful web version of this article.) (Color figure online)

4   Conclusion

In this work, we proposed a cascade multi-modality pathology segmentation framework. It has been evaluated on scar and edema segmentation of CMR images. The experimental results show our CAFB is capable in fusing complementary information from multi-sequence CMRs to boost the pathology segmentation performance. Besides, we present the advantage of using DAE to reconstruct the segmentation result with plausible shape. Future research aims to investigate the performance of the framework on other pathological datasets.

References

1. Albà, X., Figueras i Ventura, R.M., Lekadir, K., Frangi, A.F.: Healthy and scar myocardial tissue classification in DE-MRI. In: Camara, O., Mansi, T., Pop, M., Rhode, K., Sermesant, M., Young, A. (eds.) STACOM 2012. LNCS, vol. 7746, pp. 62–70. Springer, Heidelberg (2013). https://doi.org/10.1007/978-3-642-36961-2_8
2. Amado, L.C., et al.: Accurate and objective infarct sizing by contrast-enhanced magnetic resonance imaging in a canine myocardial infarction model. J. Am. Coll. Cardiol. **44**(12), 2383–2389 (2004)
3. Berry, C., et al.: Magnetic resonance imaging delineates the ischemic area at risk and myocardial salvage in patients with acute myocardial infarction. Circ. Cardiovasc. Imaging **3**(5), 527–535 (2010)
4. Hu, J., Shen, L., Sun, G.: Squeeze-and-excitation networks. In: Proceedings of the IEEE Conference on Computer Vision and Pattern Recognition, pp. 7132–7141 (2018)
5. Ingkanisorn, W.P., Rhoads, K.L., Aletras, A.H., Kellman, P., Arai, A.E.: Gadolinium delayed enhancement cardiovascular magnetic resonance correlates with clinical measures of myocardial infarction. J. Am. Coll. Cardiol. **43**(12), 2253–2259 (2004)
6. Kolipaka, A., Chatzimavroudis, G.P., White, R.D., O'Donnell, T.P., Setser, R.M.: Segmentation of non-viable myocardium in delayed enhancement magnetic resonance images. Int. J. Cardiovasc. Imaging **21**(2–3), 303–311 (2005)
7. Li, L., Weng, X., Schnabel, J.A., Zhuang, X.: Joint left atrial segmentation and scar quantification based on a DNN with spatial encoding and shape attention. arXiv preprint arXiv:2006.13011 (2020)
8. Li, L., et al.: Atrial scar quantification via multi-scale CNN in the graph-cuts framework. Med. Image Anal. **60**, 101595 (2020)
9. Spiewak, M., et al.: Comparison of different quantification methods of late gadolinium enhancement in patients with hypertrophic cardiomyopathy. Eur. J. Radiol. **74**(3), e149–e153 (2010)
10. Wang, G., Li, W., Ourselin, S., Vercauteren, T.: Automatic brain tumor segmentation using cascaded anisotropic convolutional neural networks. In: Crimi, A., Bakas, S., Kuijf, H., Menze, B., Reyes, M. (eds.) BrainLes 2017. LNCS, vol. 10670, pp. 178–190. Springer, Cham (2018). https://doi.org/10.1007/978-3-319-75238-9_16
11. Yang, G., et al.: Fully automatic segmentation and objective assessment of atrial scars for long-standing persistent atrial fibrillation patients using late gadolinium-enhanced mri. Med. Phys. **45**(4), 1562–1576 (2018)
12. Yue, Q., Luo, X., Ye, Q., Xu, L., Zhuang, X.: Cardiac segmentation from LGE MRI using deep neural network incorporating shape and spatial priors. In: Shen, D., et al. (eds.) MICCAI 2019. LNCS, vol. 11765, pp. 559–567. Springer, Cham (2019). https://doi.org/10.1007/978-3-030-32245-8_62
13. Zabihollahy, F., Rajchl, M., White, J.A., Ukwatta, E.: Fully automated segmentation of left ventricular scar from 3D late gadolinium enhancement magnetic resonance imaging using a cascaded multi-planar U-net (CMPU-Net). Med. Phys. **47**(4), 1645–1655 (2020)
14. Zabihollahy, F., White, J.A., Ukwatta, E.: Convolutional neural network-based approach for segmentation of left ventricle myocardial scar from 3D late gadolinium enhancement MR images. Med. Phys. **46**(4), 1740–1751 (2019)

15. Zeng, Q., et al.: Liver segmentation in magnetic resonance imaging via mean shape fitting with fully convolutional neural networks. In: Shen, D., et al. (eds.) MICCAI 2019. LNCS, vol. 11765, pp. 246–254. Springer, Cham (2019). https://doi.org/10.1007/978-3-030-32245-8_28

16. Zhang, L., et al.: Myocardial infarct sizing by late gadolinium-enhanced MRI: Comparison of manual, full-width at half-maximum, and N-standard deviation methods. J. Magn. Reson. Imaging **44**(5), 1206–1217 (2016)

17. Zhou, T., Fu, H., Chen, G., Shen, J., Shao, L.: Hi-net: hybrid-fusion network for multi-modal MR image synthesis. IEEE Trans. Med. Imaging **39**(9), 2772–2781 (2020)

18. Zhou, T., Ruan, S., Canu, S.: A review: deep learning for medical image segmentation using multi-modality fusion. Array **3**, 100004 (2019)

19. Zhuang, X.: Multivariate mixture model for cardiac segmentation from multi-sequence MRI. In: Ourselin, S., Joskowicz, L., Sabuncu, M.R., Unal, G., Wells, W. (eds.) MICCAI 2016. LNCS, vol. 9901, pp. 581–588. Springer, Cham (2016). https://doi.org/10.1007/978-3-319-46723-8_67

20. Zhuang, X.: Multivariate mixture model for myocardial segmentation combining multi-source images. IEEE Trans. Pattern Anal. Mach. Intell. **41**(12), 2933–2946 (2019)

21. Zhuang, X., et al.: Cardiac segmentation on late gadolinium enhancement MRI: a benchmark study from multi-sequence cardiac MR segmentation challenge. arXiv preprint arXiv:2006.12434 (2020)

Myocardial Edema and Scar Segmentation Using a Coarse-to-Fine Framework with Weighted Ensemble

Shuwei Zhai, Ran Gu, Wenhui Lei, and Guotai Wang[✉]

School of Mechanical and Electrical Engineering, University of Electronic Science
and Technology of China, Chengdu, China
guotai.wang@uestc.edu.cn

Abstract. In this work, we implement a deep learning-based segmentation algorithm that can automatically segment left ventricular (LV) blood pool, right ventricular (RV) blood pool, LV normal myocardium, LV myocardial edema and LV myocardial scar from multi-sequence Cardiac Magnetic Resonance (CMR) images. Since the edema and scar region is very small, we adapt a coarse-to-fine segmentation strategy that contains two segmentation neural networks. Firstly, we use a coarse segmentation model to predict the cardiac structure area especially the myocardium part where the scar and edema regions distribute. Then we use the fine segmentation model to get a detailed prediction for edema and scar regions. Finally, we apply a weighted ensemble model to integrate the prediction from 2D and 2.5D networks. Our proposed framework achieves an average Dice score of 0.64 for LV myocardial scar and 0.41 for LV myocardial edema on 5-fold cross validation dataset from myocardial pathology segmentation combining multi-sequence CMR(MyoPS) challenge, while achieving an average Dice score of 0.67 and 0.73 in LV myocardial scar and the union of scar and edema on test set, respectively.

Keywords: Multi-sequence CMR segmentation · Coarse-to-fine · Weighted ensemble

1 Introduction

Assessment of myocardial viability is essential in the diagnosis and treatment management for patients suffering from myocardial disease [13]. Cardiac magnetic resonance (CMR) is particularly used to provide imaging anatomical and functional information of heart, such as the late gadolinium enhancement (LGE) CMR sequence which visualizes myocardial infarction, the T2-weighted CMR which images the acute injury and ischemic regions, and the balanced Steady State Free Precession (bSSFP) cine sequence which captures cardiac motions

---

S. Zhai and R. Gu—Equal contribution.

© Springer Nature Switzerland AG 2020
X. Zhuang and L. Li (Eds.): MyoPS 2020, LNCS 12554, pp. 49–59, 2020.
https://doi.org/10.1007/978-3-030-65651-5_5

and presents clear boundaries. Hence, cardiac segmentation combining multi-sequence (CMR) which highlights myocardial scar tissue is of great clinical importance, enabling quantitative measurements useful for treatment planning and patient management.

In clinical practice, the myocardium will be classified into normal, scar and edema regions, which is important for the diagnosis and treatment management of patients. This is still an arduous task because manual delineation is generally time-consuming, tedious and subject to inter- and intra-observer variations. Therefore, it is highly desirable to develop an automatic segmentation method in clinical practice. However, segmenting normal region, scar and edema regions is a challenging task due to the fact that 1) the scar and edema regions occupy very tiny area with various locations distributed on the myocardium. 2) the scar and edema regions are hard to be distinguished in all three sequences images [12]. As shown in Fig. 1, the border between myocardial scar and normal regions is blurry on balanced steady-state free precession (bSSFP) and T2 CMR images, while reluctantly can be delineated on LGE images. Moreover, the region of myocardial edema is hard to be distinguished in T2 and LGE images, while is barely seen in bSSFP images.

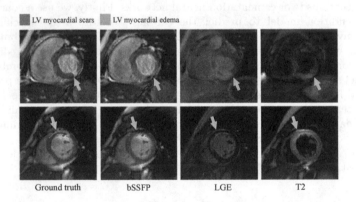

Fig. 1. Visualization comparison of LV myocardial edema and LV myocardial scar on three sequences of CMR images. The first column is the bSSFP CMR images overlapped with annotations that blue area refers to edema and the red area refers to scar. The green arrows point to the myocardial edema in three sequences of CMR images. (Color figure online)

In the literature, myocardial scar and edema segmentation tasks are under-studied until now. Eranga et al. [7] set a continuous max-flow (CMF) optimization problem to tackle myocardial infarction segmentation. Compared with traditional algorithms, deep convolutional neural networks (CNN) can automatically detects the important features without any human supervision and is also computationally efficient, which mostly are end-to-end methods to directly gain the final results [5]. Recently, a deep neural network method was proposed to segment the three cardiac structures (LV blood pool, RV blood pool, and myocardium) from LGE images, which achieve superior performance [10]. Zabihollahy et al. [11] introduced a CNN-based algorithm to obtain scar segmentation from LGE CMR images. Li et al. used a fully automated method based on the graph-cuts framework [3] and a DNN with spatial encoding and shape attention [4] to segment the left atrial and quantify left atrial scar. However, none of the above methods is designed for LV edema and scar segmentation which is tricky due to the small size and fuzzy border. Hence, it is essential to design an automatic algorithm to segment scar and edema regions on myocardium for assessment of myocardial viability.

In this paper, we propose a deep neural network-based method to perform the automatic segmentation of myocardial edema and scar by using multi-sequence CMR images. We present a coarse-to-fine framework, where we first segment the myocardium and left and right ventricular to get prior location information for the small scar and edema regions, and then we introduce myocardial prior location information to get more detailed segmentation of scar and edema regions. Our framework mainly consists of two steps:

1) A coarse segmentation network: this network is used for segmenting three cardiac structures including LV, RV and LV myocardium to get cardiac region prediction.
2) A fine segmentation network: this network is used for segmenting myocardial edema and scar regions. The input of this network is a concatenation of three-sequence CMR images and the predictions of LV, RV and LV myocardium.

There are actually two different architectures at each stage, 2D and 2.5D networks, which are of importance to segmentation of different myocardial pathology. Ensembling model of 2D and 2.5D networks can take advantage of their own learned features and boost the performance significantly. Especially we employ deep supervision for fine segmentation to strengthen connection between loss functions and feature maps at different scales.

The main contributions of our work are the following: 1) we employ a coarse-to-fine framework to address small targets with varied shapes and positions. We first predict the cardiac structure areas and serves as a location prior information to guide the myocardial scar and edema segmentation. 2) we introduce a novel ensemble method that train and predict the scar and edema regions in 2 and 2.5 dimensions. Counting for the performance in the two models, we introduce two weights to better ensemble models' performance that Dice score for LV myocardial scar and the whole region of edema and scar on test dataset are 0.67 and 0.73 respectively.

2   Method

Our proposed framework for myocardial edema and scar segmentation can be summarised as data prepossession, coarse-to-fine segmentation, and model ensemble. For data prepossession, we crop the original images with cardiac bounding box. Because it is hard to segment the myocardial edema and scar directly, we split this task into two steps: a coarse step of LV myocardium segmentation and a fine step of edema and scar segmentation, as shown in Fig. 2. For model ensemble, a weighted ensemble method is exploited to get better prediction for each classes.

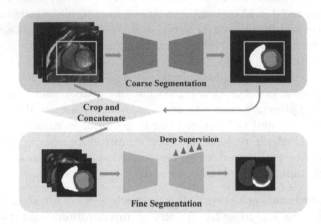

Fig. 2. Overview of coarse-to-fine segmentation network. Both loss functions of two segmentation networks are sum of cross entropy loss and Dice loss.

2.1 Data Preprocessing

In coarse segmentation, due to the imbalance of foreground and background, we count the coordinate range of the targets in the training data. We expanded 30 voxels along each dimension, and crop the images accordingly (see Fig. 3). In the fine segmentation, we similarly crop out the targets based on coarse segmentation results to reduce false positive.

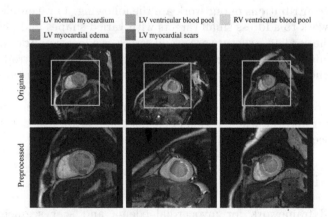

Fig. 3. Examples of bSSFP images overlapped with ground truth of training set. The first row are original images, and the second row are images after preprocessing which are cropped out with green bounding boxes. (Color figure online)

2.2 Coarse Segmentation Network

From Fig. 3, we can see that edema, scar and LV normal myocardium form complete ring-shaped myocardium that can be used as shape prior information for edema and scar. In coarse segmentation, we train a vanilla U-Net to segment 4 categories (i.e. background, complete ring-shaped myocardium, left ventricular (LV) blood pool and right ventricular (RV) blood pool). The input data set contains three sequences of CMR images: bSSFP, LGE and T2 which we directly concatenate to form 3-channel input. We use cross entropy loss and Dice loss for coarse segmentation training.

2.3 Fine Segmentation Network

In the fine segmentation, the input images are cropped according to coarse predictions, and then we concatenate the coarse segmentation map and the cropped images to form a 4-channel input to feed into the fine segmentation network. Our fine segmentation network's architecture is a variant of U-Net [6], as presented in Fig. 4, where we add instance normalization (IN) and leaky rectified linear unit (Leaky ReLU) following every convolution layer.

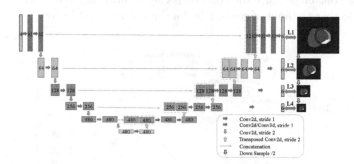

Fig. 4. Fine segmentation network architecture. When pink arrows are implemented as 2D convolutions, the network is a 2D U-Net, and when they are implemented as 3D convolutions, the network is a 2.5D U-Net. (Color figure online)

2.5D U-Net can encode the in-plane and through-plane information while 2D U-Net will ignore through-plane correlation [9], so 2.5D U-Net deal with CMR images with anisotropic resolution better than 2D U-Net.

For network training, we employ deep supervision by downsampling ground truth to 4 scales to supervise correlated scale feature maps [2]. The fine segmentation network combines deep supervision which acts as a regularization to overcome overfitting on this small data set. The overall loss function is formulated as:

$$L = \frac{1L_1 + 0.5L_2 + 0.25L_3 + 0.125L_4}{1 + 0.5 + 0.25 + 0.125},$$

where L_1, L_2, L_3 and L_4 are loss functions (i.e. sum of cross entropy loss and Dice loss) for multiple scales. By combining the cropped images that contains texture information and the coarse segmentation that contains shape prior information, the coarse-to-fine network can produce more accurate segmentation.

2.4   Weighted Ensemble

To sufficiently exploit 2D and 2.5D networks' performance advantage, we use a weighted ensemble strategy as shown in Fig. 5, where the prediction maps have 6 channels, and each channel represents one segmentation target. Let P_c^{output}, P_c^{2D} and $P_c^{2.5D}$ denote the channel c of output maps, the channel c of prediction maps of 2D U-Net and 2.5D U-Net respectively. Let w_c^{2D} and $w_c^{2.5D}$ denote the weights of channel c of 2D U-Net and 2.5D U-Net. Our weighted ensemble method can be expressed as:

$$P_c^{output} = w_c^{2D} P_c^{2D} + w_c^{2.5D} P_c^{2.5D}, c \in \{1, 2, 3, 4, 5, 6\}.$$

In Fig. 5, the weights for edema($w_5^{2D} = 0.8$, $w_5^{2.5D} = 0.2$) and scar channels($w_6^{2D} = 0.2$, $w_6^{2.5D} = 0.8$) are different while the weights for other channels are the same. So the edema channel of output is dominated by 2D U-Net that has better performance on edema segmentation than 2.5D U-Net, and similarly the scar channel of output is dominated by 2.5D U-Net.

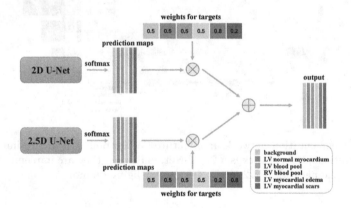

Fig. 5. Weighted ensemble strategy that we use different weights for the channels representing edema and scar of 2D U-Net and 2.5D U-Net, where different color represents different segmentation targets.

3   Experiments and Results

3.1   Data and Implementation

There are 45 cases of multi-sequence CMR (25 cases for training and 20 cases for testing), each of which refers to a patient with three sequence CMR, i.e., LGE, T2 and bSSFP CMR. The data have been pre-processed using the MvMM method [12,13], to align the three-sequence CMR into a common space and to resample them into the same spatial resolution. In our experiments we used 5-fold cross validation and calculated the average Dice score, 20 cases for training and 5 cases for validation in each fold. Dice score for every case is defined as follow:

$$Dice_c = \frac{2\sum_i P_{c,i}{}^{one-hot} G_{c,i}{}^{one-hot} + \epsilon}{\sum_i (P_{c,i}{}^{one-hot} + G_{c,i}{}^{one-hot}) + \epsilon}, c \in \{1,2,3,4,5,6\},$$

where $P_{c,i}{}^{one-hot}$ and $G_{c,i}{}^{one-hot}$ denote the values for voxel i of channel c of one-hot output and ground truth, and ϵ is a small number for numerical stability. We implemented the coarse segmentation network by PyMIC, a PyTorch library provided for medical image segmentation[1] [8], and trained the model on an NVIDIA GeForce RTX 2080 Ti with 11 GB RAM. The data augmentation before fine segmentation network was implemented using nnU-Net[2] [1].

3.2   Results

Loss Functions. We first compared several loss functions for the 2D fine segmentation network and the results are presented in Table 1.

Table 1. Dice scores of different loss functions on 5-fold validation dataset.

Loss	Dice		
	Edema	scar	Average
Dice Loss	0.3109	**0.6140**	0.4625
Cross Entropy Loss	0.3974	0.6053	0.5014
CE Loss + Dice Loss	**0.4124**	0.6071	**0.5098**

We can see that Dice loss is not as good as cross entropy loss for this task, especially for edema due to its small size. Combining the two loss functions can improve the performance obviously, which increase the edema segmentation Dice from 0.31 to 0.41 compared with only using Dice loss. The average Dice for the edema and scar is 0.51.

Coarse-to-Fine Strategy. In order to verify the effectiveness of our coarse-to-fine method, we compared it with the one-stage segmentation method (i.e., directly segmenting the edema and scar). Besides, we conducted ablation

---

[1] https://github.com/HiLab-git/PyMIC.
[2] https://github.com/MIC-DKFZ/nnU-Net.

experiments on deep supervision and 2.5D networks, and the result is shown in Table 2. Finally, we adopt weighted ensemble to integrate advantages of two models. Qualitative results of our proposed method are shown in Fig. 6.

Table 2. Experimental results on validation set with different strategies. Coarse-to-fine means using two-stage network. Coarse Map means concatenating coarse predicted map and multi-sequence images before feed into fine segmentation network. DS means deep supervision.

Coarse-to-fine	Coarse Map	DS	Network	Dice		
				Edema	scar	Average
		✓	2D	0.3880	0.5926	0.4903
✓		✓	2D	0.3870	0.6237	0.5054
✓	✓		2D	0.3798	0.6016	0.4907
✓	✓	✓	2D	**0.4124**	0.6071	**0.5098**
✓	✓	✓	2.5D	0.3602	**0.6296**	0.4949

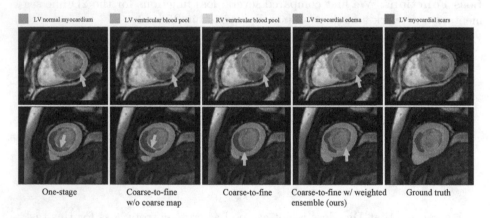

Fig. 6. Two segmentation examples as obtained by using different training combination, showing the improvement by integrating coarse-to-fine strategy (column two), coarse map concatenation feeding into fine segmentation network (column three) and weighted ensemble model (column four). The green arrows point where the prediction is quite different from ground truth. (Color figure online)

From the first and second rows of Table 2, we find that coarse-to-fine framework significantly improves small target segmentation performance, since the Dice score for scar is improved from 59.3% to 62.4%. Furthermore, from row 2 and row 4 of Table 2, we find that when feeding the coarse predicted map into fine segmentation network, the Dice score for edema can achieve at 41.2%, increasing over 2% and the average Dice can get the highest at 51.0%. Row 3 and 4 shows that introducing deep supervision can also improve model performance.

Then, we tried weighted ensemble as mentioned in Sect. 2.4. We explored 6 groups of weight value in our experiment and chose the best one as our ensemble strategy. Simple ensemble (equal weights) is better than using a single model, as shown in the first three rows in Table 3. When using different weights, the segmentation accuracy is improved by ~0.4% on our 5-fold validation dataset. Finally, we implemented ablation studies on test data as shown in Table 4, from which we can see that our method boosts the performance compared to single 2D U-Net or 2.5D U-Net. Dice score for scar is 67.2% on test dataset, much higher than 63.8% on validation dataset, which implies the uneven distribution of validation and test dataset. And it may be the reason why our method has almost the same Dice score as simple ensemble on test dataset.

Table 3. Average Dice scores on 5-fold validation dataset with different ensemble strategies. First two rows are results without ensemble.

Edema weights (2D, 2.5D)	Scar weights (2D, 2.5D)	Dice		
		Edema	scar	Average
2D U-Net		0.4124	0.6070	0.5098
2.5D U-Net		0.3602	0.6292	0.4949
(0.5, 0.5)	(0.5, 0.5)	0.4049	0.6370	0.5210
(0.6, 0.4)	(0.4, 0.6)	0.4099	0.6391	0.5245
(0.7, 0.3)	(0.3, 0.7)	0.4120	**0.6383**	0.5252
(0.8, 0.2)	**(0.2, 0.8)**	**0.4130**	0.6376	**0.5253**
(0.9, 0.1)	(0.1, 0.9)	0.4122	0.6362	0.5242
(1.0, 0.0)	(0.0, 1.0)	0.4121	0.6354	0.5238

Table 4. Average Dice scores on test dataset.

Network	Dice		
	scar	scar + Edema	Average
2D U-Net	0.6409	0.6954	0.6682
2.5D U-Net	0.6614	0.7294	0.6954
Simple ensemble	0.6716	**0.7322**	0.7019
Ours	**0.6723**	0.7314	**0.7019**

4   Discussion and Conclusion

In this paper, we proposed a coarse-to-fine framework to segment myocardial edema and scar that are very small targets. The coarse-to-fine framework contains two stages. The coarse segmentation framework is to predict three cardiac structures (LV, RV and myocardium) to get an approximate position of two small regions because myocardial edema and scar are both distributed on myocardium. The fine segmentation framework of which input is concatenation of coarse network output prediction serving as prior location information and three sequences of CMR images to conduct detailed target prediction. Meanwhile, we used deep supervision to strengthen supervision of each scale feature maps because the feature information related to our two small targets at each scale are very important for segmentation. We also introduced a novel weighted ensemble method that gives a specific weight to 2D and 2.5D fine segmentation network according to the model's performance in each class. Our coarse-to-fine framework shows a great performance on test set, and can extend to other tasks for small target segmentation.

References

1. Isensee, F., et al.: nnU-Net: self-adapting framework for U-net-based medical image segmentation. arXiv preprint arXiv:1809.10486 (2018)
2. Lee, C.Y., Xie, S., Gallagher, P., Zhang, Z., Tu, Z.: Deeply-supervised nets. In: Artificial Intelligence and Statistics, pp. 562–570 (2015)
3. Li, L., Weng, X., Schnabel, J.A., Zhuang, X.: Joint left atrial segmentation and scar quantification based on a DNN with spatial encoding and shape attention. arXiv preprint arXiv:2006.13011 (2020)
4. Li, L., et al.: Atrial scar quantification via multi-scale CNN in the graph-cuts framework. Med. Image Anal. **60**, 101595 (2020)
5. Long, J., Shelhamer, E., Darrell, T.: Fully convolutional networks for semantic segmentation. In: Proceedings of the IEEE Conference on Computer Vision and Pattern Recognition, pp. 3431–3440 (2015)
6. Ronneberger, O., Fischer, P., Brox, T.: U-net: convolutional networks for biomedical image segmentation. In: Navab, N., Hornegger, J., Wells, W.M., Frangi, A.F. (eds.) MICCAI 2015. LNCS, vol. 9351, pp. 234–241. Springer, Cham (2015). https://doi.org/10.1007/978-3-319-24574-4_28
7. Ukwatta, E., et al.: Myocardial infarct segmentation from magnetic resonance images for personalized modeling of cardiac electrophysiology. IEEE Trans. Med. Imaging **35**(6), 1408–1419 (2015)
8. Wang, G., et al.: A noise-robust framework for automatic segmentation of Covid-19 pneumonia lesions from CT images. IEEE Trans. Med. Imaging **39**(8), 2653–2663 (2020)
9. Wang, G., et al.: Automatic segmentation of vestibular Schwannoma from T2-weighted MRI by deep spatial attention with hardness-weighted loss. In: Shen, D., et al. (eds.) MICCAI 2019. LNCS, vol. 11765, pp. 264–272. Springer, Cham (2019). https://doi.org/10.1007/978-3-030-32245-8_30

10. Yue, Q., Luo, X., Ye, Q., Xu, L., Zhuang, X.: Cardiac segmentation from LGE
    MRI using deep neural network incorporating shape and spatial priors. In: Shen,
    D., et al. (eds.) MICCAI 2019. LNCS, vol. 11765, pp. 559–567. Springer, Cham
    (2019). https://doi.org/10.1007/978-3-030-32245-8_62
11. Zabihollahy, F., White, J.A., Ukwatta, E.: Convolutional neural network-based
    approach for segmentation of left ventricle myocardial scar from 3D late gadolinium
    enhancement MR images. Med. Phys. 46(4), 1740–1751 (2019)
12. Zhuang, X.: Multivariate mixture model for cardiac segmentation from multi-
    sequence MRI. In: Ourselin, S., Joskowicz, L., Sabuncu, M.R., Unal, G., Wells,
    W. (eds.) MICCAI 2016. LNCS, vol. 9901, pp. 581–588. Springer, Cham (2016).
    https://doi.org/10.1007/978-3-319-46723-8_67
13. Zhuang, X.: Multivariate mixture model for myocardial segmentation combining
    multi-source images. IEEE Trans. Pattern Anal. Mach. Intell. 41(12), 2933–2946
    (2019)

Exploring Ensemble Applications for Multi-sequence Myocardial Pathology Segmentation

Markus J. Ankenbrand$^{(\boxtimes)}$ (iD), David Lohr (iD), and Laura M. Schreiber (iD)

Chair of Cellular and Molecular Imaging, Comprehensive Heart Failure Center,
University Hospital Würzburg, Am Schwarzenberg 15, 97078 Würzburg, Germany
Ankenbrand_M@ukw.de

Abstract. We tested different loss functions and hyper-parameters using a 2D U-Net architecture (resnet34 backbone) with five-fold cross-validation on the training data. Pathology specific sequence data (e.g. LGE for scar and T_2 for edema) was used as a sole input for training and in combination with all sequences. We wanted to address the question whether for limited training data it is beneficial to incorporate prior knowledge by predicting classes with their appropriate sequence or if a neural network is able to infer these relationships from a multi-sequence dataset. In addition, we aimed to create a model zoo, combining predictions from models with high performance on individual classes. Images were cropped to the central 256×256 region as this contained the region of interest in all cases. To improve robustness and learn more general features extensive data augmentation was used, including both MR artifacts (motion, noise) and standard image transformations (zoom, rotation, brightness, contrast). Variations of training data, loss functions and hyper-parameters led to 21 models trained. The multi-sequence model was trained using all image sequences input via color channels producing pixel-level segmentation for all six classes (background, left ventricle, right ventricle, myocardium, edema, and scar). Cross-entropy as a loss function performed best (metric: dice) for non-pathologic tissue, while pathology weighted focal-loss (0.35 for both scar and edema) had best mean performance on scar and edema.

These results indicate that the employed neural networks are capable of learning multi-sequence segmentation end-to-end. Combining different outputs from a model zoo further improved segmentation performance.

Keywords: Deep learning · U-Net · Ensemble · Segmentation · Cardiac MRI

1 Introduction

1.1 Background

Cardiac magnetic resonance (CMR) imaging applies methods to investigate cardiac function and pathologies non-invasively. Different measurement sequences are used to produce images with different contrast, enabling diagnosis of varying pathologic tissue alterations. It is common practice to segment the left ventricle and the myocardium to

© Springer Nature Switzerland AG 2020
X. Zhuang and L. Li (Eds.): MyoPS 2020, LNCS 12554, pp. 60–67, 2020.
https://doi.org/10.1007/978-3-030-65651-5_6

assess clinically relevant parameters like ejection fraction, stroke volume and myocardial mass as well as wall motion. Scar volume, as a result of acute myocardial infarction, has significant prognostic value for outcome prediction and treatment, thus, increasing the importance of accurate pathology segmentation. In clinical practice, such segmentations are commonly done semi-automatically. Fully automatic segmentation algorithms have been proposed using different methods, including artificial neural networks [1, 2]. However, these networks are usually trained on a single sequence and a subset of tissue/pathology classes. For the prediction of multiple pathologic tissue alterations in parallel, it might be beneficial to train segmentation networks, which combine information from multiple sequences.

1.2 Related Work

Neural networks, particularly convolutional neural networks and U-Nets [3] have been used for segmentation of cardiac magnetic resonance images [1, 4]. Beside healthy tissue also pathological classes like left ventricular scar [5] or left atrial scar have been addressed [6, 7]. However, the simultaneous use of multiple sequences and multiple classes presents a new set of challenges.

2 Experiments

In the MyoPS 2020 challenge, three different sequences (bSSFP, LGE, SPAIR) were measured for each of 45 patients, providing ground truth segmentation for left ventricular (LV) blood pool, right ventricular (RV) blood pool, LV myocardium (MY), edema, and scar for 25 patients. All data was provided aligned (MvMM method [8, 9]) in a common space with identical spatial resolution by the organizers. The aim of the challenge was to create an algorithm for pixel-wise segmentation of the pathology classes edema and scar. In this study, we employed variations of individual neural networks as well as a model ensemble, combining models with high performance on individual morphologic classes.

The experiments and parameter search were done in Google Colab GPU instances. For the final training and prediction, we used our local HPC i. 8x Intel(R) Xeon(R) CPU E5-2630 v3 @ 2.40 GHz ii. 512 GB of memory iii. 1x NVIDIA Tesla K80 with 12 GB of memory.

3 Methods

3.1 Software

We built our model using open source software including python 3.7.7, pytorch 1.5.1 [10], fastai2 0.0.17 [11], torchio 0.15.5 [12], MONAI 0.2.0, nibabel 3.0.1 [13] as well as their dependencies. Our model and code is openly available on GitHub and zenodo (code: https://github.com/chfc-cmi/miccai2020-myops and https://doi.org/10.5281/zenodo.3982324; models: https://doi.org/10.5281/zenodo.3985837).

3.2 Processing Pipeline and Architecture

We converted all images from nifti to png format saving each slice as one image with sequences combined as color channels. Additionally, each sequence was saved independently as a grey-scale image. We tried normalization of the LGE and T_2 images using contrast limited adaptive histogram equalization (CLAHE) [14]. In this step, in addition to the original images, transformed images with simulated MR artifacts (motion and noise) were produced using torchio [12]. These images were used to train U-Nets [3] with a resnet34 [15] backbone (initialized using ImageNet [16] weights) with further augmentations (rotation, brightness, contrast) with fastai2 [4, 11]. Performance of different hyper-parameter settings were evaluated using dice scores from five-fold cross-validation. The same split was used for all experiments and every data set was part of the validation set at least once.

3.3 Hyper-parameter Search

Preliminary Experiments. In preliminary experiments the effect of contrast enhancements using CLAHE as well as cropping vs resizing to 256×256 pixels were tested.

Systematic Experiments. For the general multi-channel/multi-class networks, different losses were tested. Cross-entropy loss (ce) was compared to differently weighted focal loss [17]. We experimented with some classes receiving higher weights (values used are indicated in parentheses), while the other classes received balanced weights:

- all classes with equal weights (balanced)
- myocardium (0.2, 0.3), edema (0.2, 0.3) and scar (0.2, 0.3), label: multi_pathoMyo
- edema (0.2, 0.35, 0.49) and scar (0.2, 0.35, 0.49), label: multi_patho
- edema (0.2, 0.4, 0.6, 0.8, 0.99), label: multi_edema
- scar (0.2, 0.4, 0.6, 0.8, 0.99), label: multi_scar

Additionally, pathology specific networks (t2_edemaOnly, lge_scarOnly) were trained on their corresponding sequence only (edema with T_2 and scar with LGE) using two different weightings of the focal loss (0.5 and 0.8). In total 21 networks were trained this way for 30 epochs (10 epochs frozen, 20 epochs unfrozen) and a base learning rate of 10^{-3}.

Targeted Experiments. The best performing networks from the systematic experiments were selected based on mean dice score over all cross validations. For LV, RV and myocardium only the network with the highest dice score was selected. For the pathology classes first the network with highest mean dice over both classes was selected, then for each class the two remaining networks with highest individual dice scores in the respective class were selected. This way a total of six networks were selected. These networks were trained for 60 epochs (20 frozen, 40 unfrozen) in order to assess benefits of prolonged training duration.

Final Training. For the evaluation on the test set, the six networks from the targeted experiments were trained from scratch using all 25 data sets for training and no validation set. Training was done for 60 epochs (20 frozen, 40 unfrozen), since average performance was increased with prolonged training duration.

3.4 Ensemble Method

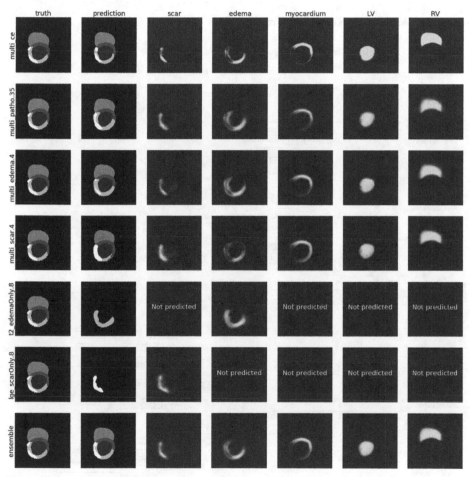

Fig. 1. Probability maps for all classes and derived prediction (second column) for the six networks and the ensemble, compared to the ground truth (first column) for a single slice of the training data. The result of the ensemble method (bottom row) is the mean over the probability maps of the six separate networks above.

The networks were trained with different foci, which led to different strengths and weaknesses. Therefore, we combined predictions from the different networks in a bagging approach. This combination included predictions from all six networks from the final training. Class probabilities were averaged over all networks, taking into account that the specialized networks only returned predictions for their respective pathology class. The final prediction for each pixel was the argmax of these averages (Fig. 1).

4   Results

4.1   Cross-Validation Results on Training Set

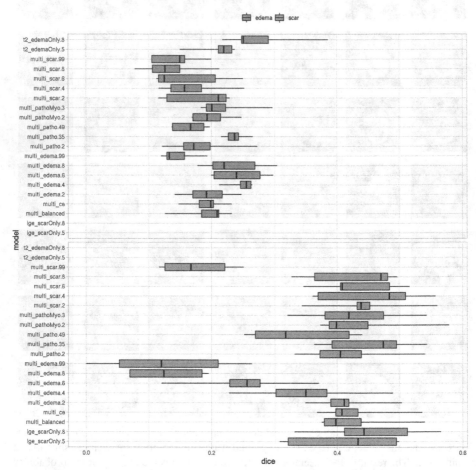

Fig. 2. Dice score for edema (top panel) and scar (bottom panel) over the five cross-validation folds of each of the 21 networks from the systematic parameter search. Naming of models: input channels (multi, lge, t2), focused classes (scar, edema, patho: scar + edema, pathoMyo: scar + edema + MY, ce for cross-entropy and balanced have identical weight for all classes) and weight for those classes as suffix.

Preliminary experiments indicated that not using CLAHE and cropping to 256 × 256 pixels yields better results than normalization or resizing. Thus, only cropping was used in the systematic experiments. In the systematic experiments, the network with cross-entropy loss reached the best results for LV, RV and myocardium segmentation with mean values of $dice_{LV} = 0.855$, $dice_{RV} = 0.783$ and $dice_{MY} = 0.696$. The best mean performance on both pathology classes: $mean(dice_{edema}, dice_{scar}) = 0.345$ was achieved using the multi-channel network (multi_patho) with weights of 0.35 for both pathology classes. Of the remaining networks the highest dice on scar was reached by the multi-channel network (multi_scar, weight: 0.4) and by the specialized LGE network (lge_scarOnly, weight 0.8) (Fig. 2), while the specialized T_2 network (t2_edemaOnly, weight: 0.8) and the multi-channel network (multi_edema, weight: 0.4) reached the highest dice scores for edema (Fig. 2). Longer training improved dice scores for almost all classes and networks (Fig. 3, Table 1).

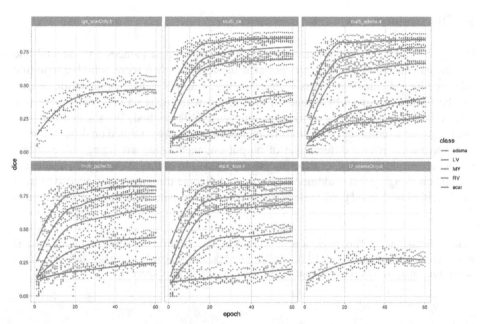

Fig. 3. Dice scores throughout training of the six networks from the targeted experiments. Data for all five cross-validation folds is shown with loess-smoothed lines for each class. The first 20 epochs were trained with frozen weights, the remaining 40 with unfrozen weights.

4.2 Performance on Test Set

Evaluation results on the test set were provided by the challenge organizers for two models, the multi_patho.35 network and the ensemble method. The ensemble reached better performance with mean ± standard deviation of $dice_{scar} = 0.620 \pm 0.240$ and $dice_{edema+scar} = 0.665 \pm 0.137$ compared to $dice_{scar} = 0.593 \pm 0.232$ and $dice_{edema+scar} = 0.611 \pm 0.111$ for the single network. For all but one patient dice scores for scar were greater than 0 indicating at least some overlap between truth and prediction.

Table 1. Mean performance of the targeted experiment networks over the five-fold cross-validation after 60 epochs of training. Highest dice for each class in bold.

Network	$dice_{LV}$	$dice_{MY}$	$dice_{RV}$	$dice_{edema}$	$dice_{scar}$	mean dice edema, scar
multi_patho.35	0.829	0.649	0.773	0.246	0.441	**0.343**
multi_scar.4	0.850	0.690	0.779	0.202	**0.479**	0.341
multi_ce	**0.853**	**0.695**	**0.787**	0.227	0.438	0.333
multi_edema.4	0.843	0.664	0.781	0.261	0.400	0.330
lge_scarOnly.8	–	–	–	–	0.467	–
t2_edemaOnly.8	–	–	–	**0.276**	–	–

5 Discussion

It is possible to train neural networks both on separate sequences and on multiple sequences with good performance. For scar the reported dice score is higher than that achieved by individual observers reported as 0.524 ± 0.158 [9]. Segmentation quality can be further improved by training a model zoo with focus on different classes and combining their predictions using a bagging ensemble method. We showed that it is even possible to combine predictions from networks that were trained on different input data (channels) with a different set of output channels using averaging. While these results are promising, further experiments are needed to optimize the hyper-parameters for this challenging task. Additionally more and diverse training data is needed to train an algorithm with good performance and to reliably estimate its performance on unseen data.

Acknowledgements. This work was supported by the German Ministry of Education and Research (grant number, 01EO1504). The funding body took no role in the design of the study, collection, analysis, and interpretation of data and in writing the manuscript.

References

1. Bai, W., et al.: Automated cardiovascular magnetic resonance image analysis with fully convolutional networks. J. Cardiovas. Magn. Resonan. **20**, 65 (2018)
2. Baumgartner, C.F., Koch, L.M., Pollefeys, M., Konukoglu, E.: An exploration of 2D and 3D deep learning techniques for cardiac MR image segmentation. In: Pop, M., et al. (eds.) STACOM 2017. LNCS, vol. 10663, pp. 111–119. Springer, Cham (2018). https://doi.org/10.1007/978-3-319-75541-0_12
3. Ronneberger, O., Fischer, P., Brox, T.: U-Net: convolutional networks for biomedical image segmentation. In: Navab, N., Hornegger, J., Wells, William M., Frangi, Alejandro F. (eds.) MICCAI 2015. LNCS, vol. 9351, pp. 234–241. Springer, Cham (2015). https://doi.org/10.1007/978-3-319-24574-4_28

4. Ankenbrand, M.J., Lohr, D., Schlötelburg, W., Reiter, T., Wech, T., Schreiber, L.M.: A Deep Learning Based Cardiac Cine Segmentation Framework for Clinicians - Transfer Learning Application to 7T. medRxiv 2020.2006.2015.20131656 (2020)
5. Zabihollahy, F., White, J.A., Ukwatta, E.: Convolutional neural network-based approach for segmentation of left ventricle myocardial scar from 3D late gadolinium enhancement MR images. Med. Phys. **46**, 1740–1751 (2019)
6. Li, L., Weng, X., Schnabel, J.A., Zhuang, X.J.A.: Joint left atrial segmentation and scar quantification based on a DNN with spatial encoding and shape attention. abs/2006.13011 (2020)
7. Li, L., et al.: Atrial scar quantification via multi-scale CNN in the graph-cuts framework. Med. Image Anal. **60**, 101595 (2020)
8. Zhuang, X.: Multivariate mixture model for cardiac segmentation from multi-sequence MRI. In: Ourselin, S., Joskowicz, L., Sabuncu, Mert R., Unal, G., Wells, W. (eds.) MICCAI 2016. LNCS, vol. 9901, pp. 581–588. Springer, Cham (2016). https://doi.org/10.1007/978-3-319-46723-8_67
9. Zhuang, X.: Multivariate mixture model for myocardial segmentation combining multi-source images. IEEE Trans. Pattern Anal. Mach. Intell. **41**(12), 2933–2946 (2019)
10. Paszke, A., et al.: PyTorch: an imperative style, high-performance deep learning library. Adv. Neural. Inf. Process. Syst. **32**, 8024–8035 (2019)
11. Howard, J., Gugger, S.: fastai: a layered api for deep learning. Information **11**, 108 (2020)
12. Pérez-García, F., Sparks, R., Ourselin, S.: TorchIO: a Python library for efficient loading, preprocessing, augmentation and patch-based sampling of medical images in deep learning. arXiv:2003.04696 [cs, eess, stat] (2020)
13. Brett, M., et al.: freec84: nibabel. Zenodo (2020). https://doi.org/10.5281/zenodo.591597
14. Pizer, S.M., et al.: Adaptive histogram equalization and its variations. Comput. Vis. Graph **39**, 355–368 (1987)
15. He, K., Zhang, X., Ren, S., Sun, J.: Deep residual learning for image recognition. In: IEEE Conference on Computer Vision and Pattern Recognition (CVPR), pp. 770–778 (2016)
16. Russakovsky, O., et al.: ImageNet large scale visual recognition challenge. Int. J. Comput. Vision **115**(3), 211–252 (2015). https://doi.org/10.1007/s11263-015-0816-y
17. Lin, T.-Y., Goyal, P., Girshick, R., He, K., Dollár, P.: Focal loss for dense object detection. arXiv:1708.02002 [cs] (2017)

Max-Fusion U-Net for Multi-modal Pathology Segmentation with Attention and Dynamic Resampling

Haochuan Jiang[1], Chengjia Wang[2(✉)], Agisilaos Chartsias[1], and Sotirios A. Tsaftaris[1,3]

[1] School of Engineering, University of Edinburgh, Edinburgh, UK
[2] Center for Cardiovascular Science, University of Edinburgh, Edinburgh, UK
chengjia.wang@ed.ac.uk
[3] The Alan Turing Institute, London, UK

Abstract. Automatic segmentation of multi-sequence (multi-modal) cardiac MR (CMR) images plays a significant role in diagnosis and management for a variety of cardiac diseases. However, the performance of relevant algorithms is significantly affected by the proper fusion of the multi-modal information. Furthermore, particular diseases, such as myocardial infarction, display irregular shapes on images and occupy small regions at random locations. These facts make pathology segmentation of multi-modal CMR images a challenging task. In this paper, we present the Max-Fusion U-Net that achieves improved pathology segmentation performance given aligned multi-modal images of LGE, T2-weighted, and bSSFP modalities. Specifically, modality-specific features are extracted by dedicated encoders. Then they are fused with the pixel-wise maximum operator. Together with the corresponding encoding features, these representations are propagated to decoding layers with U-Net skip-connections. Furthermore, a spatial-attention module is applied in the last decoding layer to encourage the network to focus on those small semantically meaningful pathological regions that trigger relatively high responses by the network neurons. We also use a simple image patch extraction strategy to dynamically resample training examples with varying spacial and batch sizes. With limited GPU memory, this strategy reduces the imbalance of classes and forces the model to focus on regions around the interested pathology. It further improves segmentation accuracy and reduces the mis-classification of pathology. We evaluate our methods using the Myocardial pathology segmentation (MyoPS) combining the multi-sequence CMR dataset which involves three modalities. Extensive experiments demonstrate the effectiveness of the proposed model which outperforms the related baselines. The code is available at https://github.com/falconjhc/MFU-Net.

Keywords: Pathology segmentation · Multi-modal · Max-fusion · Dynamic resample

© Springer Nature Switzerland AG 2020
X. Zhuang and L. Li (Eds.): MyoPS 2020, LNCS 12554, pp. 68–81, 2020.
https://doi.org/10.1007/978-3-030-65651-5_7

1   Introduction

Cardiac diseases are typically assessed using multiple cardiac MR (CMR) sequences (modalities), providing complementary information. For example, Late Gadolinium Enhancement (LGE) detects myocardial infarct, T2-weighted (T2) images provide clear visibility of acute injury and ischemic regions, and balanced-Steady State Free Precession cine sequence (bSSFP) offers high contrast between anatomical regions and captures cardiac motion.

Deep learning models have been extensively used for automatic segmentation of multi-modal data. A critical step for the analysis of multi-modal CMR data is to effectively fuse information from multiple modalities. Prior works [5] concatenate the feature maps extracted from different modalities into different channels and fuse them in the following convolutional layers. Other methods [3,9] merge the features across different layers of the neural network, where a cross-modal convolution fusion model is introduced in [13]. In [7] they employ dedicated encoders for different modalities to encode different types of information, for example, content and style features from the corresponding input data. The features are then fused using channel concatenation in the U-Net skipping-connections. A similar idea was also used in [1] where a maximum fusion operator instead of simple concatenation in the skip-connections is applied on disentangled anatomy factors extracted from different modalities at the end of encoders.

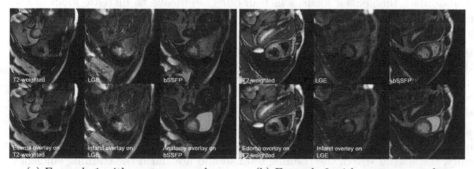

(a) Example 1 with anatomy overlay   (b) Example 2 with anatomy overlay

Fig. 1. Examples of multi-modal CMR images overlaying anatomy and pathology.

One other challenge in segmenting pathology such as myocardial infarct and edema is that these pathologies are often of diverse shape and occur at random positions. As such, shape priors such as mask discriminator [1] cannot be used. Besides, the interested pathology and anatomy only occur within a small region of the whole image, as examples of multi-sequence CMR images with manually segmented anatomy and pathology (myocardial infarction and edema) given in Fig. 1. This makes the data distribution highly imbalanced across classes, resulting in overfitting in the training data. Particularly in current popular backbone convolutional neural networks (CNN) assuming all pixels in the image contribute equally to the final prediction, the over-fitting issue is even worse. A possible solution is to use the spatial attention module [4], leading the network to focus on specific image regions. In our case, the focus corresponds to pathology pixels.

In addition, given limitations in GPU memory, training can only be performed with a small batch size. This even worsen the overfitting issue since due to this and small pathological region in each image, in each training iteration, only a small amount of pathology pixels are seen by the network. Nevertheless, the batch size can be increased if training with smaller size patches instead of full images, e.g., by engaging random cropping. Although it is commonly used as a data augmentation technique [12], all patches are treated equally importantly. It is appealing if the cropping strategy will oversample patches around pathology regions that we are interested in.

In this paper, we propose the Max-Fusion U-Net (MFU-Net) for cardiac pathology segmentation, given fully-annotated multi-modal aligned images. We use dedicated encoders to extract features for each modality, as in [1,7]. But rather than channel concatenation [7], we fuse features from different modalities with the pixel-wise maximum operator applied on each layer [1]. This fusion operator guides the network to keep informative features extracted by each modality. At the same time, fusion with maximum operator indirectly encourages feature maps to encode important features in high intensities including pathological pixels. A spatial-attention module is also employed in the last decoding layer to modulate the spatial focus, which in our case means to increase focus of the pathology pixels. Finally, to address the issue that only a small amount of pathological pixels are exposed to the network during training, we adopt a dynamic resampling strategy. To obtain each batch, we extract multiple patches around the interested pathology based on an arbitrary probability, then extract the rest data by randomly cropping the image to the same size. By feeding more patches related to pathological regions and less related to background patches, the network will thus naturally become more sensitive to pathological pixels. At the same time, the training batch size can be dynamically enlarged without occupying extra computation resources due to the reduced image dimension. Theoretically, the spatial size of the training data should not harm the training efficiency as long as the sampled image patches are bigger than the largest receptive field of the network. Extensive experiments have demonstrated the effectiveness of the proposed MFU-Net in cardiac pathology segmentation including infarction and edema when given multi-modal inputs including LGE, T2-weighted, and bSSFP, outperforming relevant methods. Major **contributions** of this work are summarized as follows:

- We proposed the MFU-Net that fuses multi-modal features extracted by dedicated encoders with the pixel-wise maximum operator;
- We incorporate a spatial-attention module to guide the network to focus on the pathology region;

- We proposed a novel training strategy by feeding randomly resampled sub-patches from the original training data with more probability around the pathology region, at the same time increasing the batch size dynamically;
- MFU-Net improves the Dice score of state-of-the-art benchmarks on myocardial pathology segmentation on multi-modal CMR 2020 dataset [15,16].

2 Methodology

This section presents the proposed MFU-Net model, and the details about the architecture, the modality-specific encoders, the maximum fusion operator, and the attention-based decoding modules.

Overview: Let $X_{LGE}, X_{T2}, X_{bSSFP}$ represent images of LGE, T2-weighted, and bSSFP CMR modalities respectively, and Y_{ana}, Y_{pat} be the associated anatomy and pathology masks. If i enumerates all samples from the above sets, we assume a fully labelled multi-modal pathology subset $\mathbf{L} = \{x^i_{LGE}, x^i_{T2}, x^i_{bSSFP}, y^i_{ana}, y^i_{pat}\}$, where three modality slices x^i_{LGE}, x^i_{T2}, $x^i_{bSSFP} \subset \mathbb{R}^{H \times W}$ are preprocessed [15,16], such that they are aligned in a common space and are resampled to the same spatial resolution. In addition, $y^i_{ana} \in Y_{ana} := \{0,1\}^{H \times W \times N}$, and $y^i_{pat} \in Y_{pat} := \{0,1\}^{H \times W \times K}$, where N and K denote the number of anatomy, and pathology masks respectively[1], and H and W are the image height and width.

2.1 Model Architecture

The architecture of MFU-Net is illustrated in Fig. 2. It consists of three modality-specific encoders, a multi-modal feature fusion with pixel-wise maximum operator, and a decoder with a spatial attention module that produces the segmentation results.

Individual Encoders: The original U-Net architecture [10] only specifies a single encoder to extract features. To accommodate differences in the pixel intensity distributions between modalities, we expand the U-Net by using one independent encoder for each modality. This leads to three modality-specific encoders. Represented by red, green, and blue colors in Fig. 2, these encoders are denoted as Enc_{LGE}, Enc_{T2}, and Enc_{bSSFP} respectively for LGE, T2, and bSSFP data. The encoded features $Enc_{LGE}(x^i_{LGE})$, $Enc_{T2}(x^i_{T2})$, and $Enc_{bSSFP}(x^i_{bSSFP})$ are concatenated and used as input to the bottleneck blocks (the transparent brown blocks in Fig. 2).

Modality Fusion: A simple way for feature fusion is through channel concatenation [7]. However, this strategy does not really merge the modality-specific information into modality-independent features, so that the contribution of different modalities can not be balanced dynamically. Such adaptive balancing

---

[1] We restrict to the case where $N = 3$ (myocardium, left ventricle, and right ventricle) and $K = 2$ (infarction and edema).

among modalities is particularly important in pathology segmentation, where specific pathologies can only be spot in particular modalities, i.e. infarct can only be seen in LGE, while edema can only be seen in T2, as seen in Fig. 1.

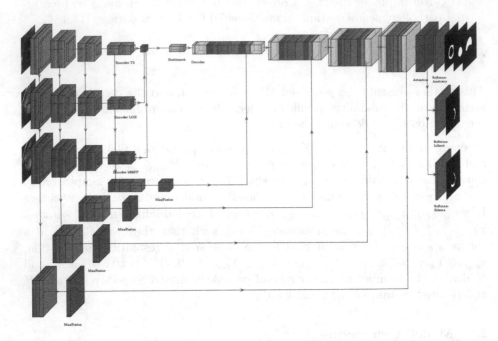

Fig. 2. MFU-Net architecture. Red, Green, and Blue blocks represent LGE, bSSFP, and T2 encoding features. Yellow blocks depict the decoding features. Pink Blocks are max-fused features, while transparent brown block is the bottleneck feature. Solid brown ones are the softmaxed probability map, while the amaranth block is the spatial attention module. (Color figure online)

Instead, we would like to fuse the feature in an auto-selective fashion. To this end, we employ the pixel-wise maximum operator, which has been previously used in [1] for dual-modal anatomy segmentation. In the proposed MFU-Net, the fusion is among features generated by the dedicated encoders, producing the fused feature as depicted in pink blocks in Fig. 2. Rather than fusing latent features of one layer [1], we apply the max-fusion operation to different blocks in the encoders for multi-scale mixture of the multi-modal information. For instance, for the k-th encoding layer, the fusion is performed by $Enc^k(x_{LGE}^i, x_{T2}^i, x_{bSSFP}^i) = \max(Enc_{LGE}^k(x_{LGE}^i), Enc_{T2}^k(x_{T2}^i), Enc_{bSSFP}^k(x_{bSSFP}^i))$ in a pixel-wise fashion.[2] It provides the dynamically selective features across modalities. However, the conventional concatenation features do not differentiate features from different modalities. The fused feature Enc^k, together with the linear concatenation of $Enc_{LGE}^k(x_{LGE}^i)$, $Enc_{T2}^k(x_{T2}^i)$, and $Enc_{bSSFP}^k(x_{bSSFP}^i)$, are then concatenated

---

[2] For simplicity we note it as Enc^k in following sections.

to the corresponding decoding layer with a skip connection, as in the original U-Net [10]. The linear concatenated and nonlinear max-fused representations provide the complementary information for the modal-specific features.

Two examples of the max-fused features compared to single-modal features are shown in Fig. 3. As discussed above, specific pathologies can only be observed in particular modalities clearly. For example, myocardial infarction can only be observed on features extracted from LGE data as a small dark area (Fig. 3c and 3d). Similarly, the boundary of edema can only be depicted on T2 feature maps (Fig. 3e and 3f). In comparison, both pathological regions can be easily detected with relatively clearer boundaries on the max-fused feature maps (Fig. 3c, 3d, 3e, 3f). Furthermore, the interested anatomical structures can be seen as easily as in bSSFP features (Fig. 3a and 3b). On the contrary, the boundaries of heart anatomy and edema are blurred in LGE, so is the infarction in T2 data. Boundaries of both infarction and edema are hard to be detected in bSSFP data. This can be seen as a qualitative evidence of an effective mixture of the multi-modality information.

Decoding with Attention: The decoder of MFU-Net receives as input the bottleneck layer that follows the concatenated multi-modal features of the encoding part. A series of convolutional blocks upsample the spatial resolution as in U-Net, and are concatenated with the encoding features (including the max-fused feature and the corresponding encoding features for each modality) computed at the corresponding layers of the encoder with skip connections.

Since cardiac pathologies often occupy in a small part of the whole image, producing segmentations by treating each pixel equally is challenging and might lead the network to concentrate more on the background but ignore tiny pathological regions. In order to overcome this issue, we use a spatial attention mechanism [4] to capture long-range pixel dependencies and assign different weights on different regions. In this sense, segmentation can be improved by selecting useful information in features extracted around the pathological regions and by discarding unrelated features. In detail, the spatial attention module, shown in Fig. 4, is applied at the last layer of the decoding path with the architecture of the spatial and channel attention modules following [4]. In order to reduce computational complexity introduced when the feature dimensions are large, we first downsample the input feature using stride-2 convolutions before calculating the query, key, and value tensors. The attention module is depicted in Fig. 4. After calculating the attention map, the dimension will be recovered by deconvolution in the upsampling block. Figure 5 gives examples of spatial attention outputs with corresponding predicted masks. Clearly, the corresponding mask region is highlighted in the spatial attention maps, demonstrating the utility of this mechanism in segmentation.

3 Implementation

In this section, the implementation details of the proposed MFU-Net will be specified. Firstly, we will introduce the dynamic resampling training strategy,

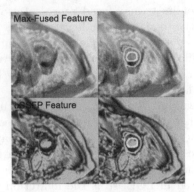

(a) Example 1 with anatomy overlay

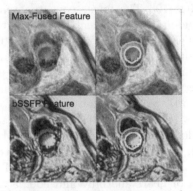

(b) Example 2 with anatomy overlay

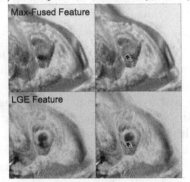

(c) Example 1 with infarct overlay

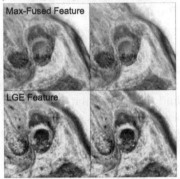

(d) Example 2 with infarct overlay

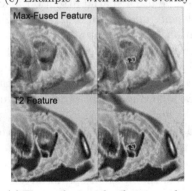

(e) Example 1 with edema overlay

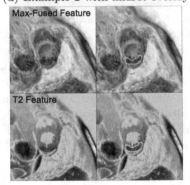

(f) Example 2 with edema overlay

Fig. 3. Two examples of comparison between the feature maps extracted before and after the max-fusion operation in terms of visibility of: (a) and (b) anatomy; (c) and (d), myocardial infarction; (e) and (f) edema. For each subfigure, the max-fused feature maps are shown at the top and modality-specific feature maps are shown at the bottom. The object boundaries overlapped with the feature maps are on the right.

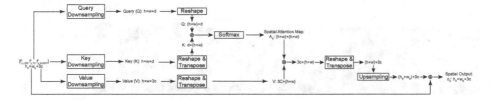

Fig. 4. Attention Module at the last decoding layer. $\oplus$ and $\otimes$ represent element-wise summation and multiplication respectively between two matrices.

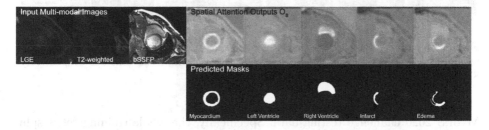

Fig. 5. Spatial attention outputs correspond to the predicted masks.

then the alternative cross-validation to make full use of the training data and avoid overfitting issue will be specified.

3.1 Dynamic Resampling Training Strategy

The proposed MFU-Net is deployed on a GTX Titan X GPU with 12GB standard memory. In the training process, the available memory allows 288×288 image size with a batch size equals to 4. In order to increase the model's focus on pathological regions, we also train with patches of different sizes that are dynamically resampled. For the batch obtained at the t-th iteration, we first decide the patch size d_t by $d_t = 96 + 16i$, $i \in \{1, \cdots, 12\}$ where i is randomly picked. Then, with an arbitrary probability ρ_c, an extracted patch is centred on the pathology of interest. The dynamic batch size N_t is decided by $N_t = \lfloor d_{t-1}^2 N_{t-1}/d_t^2 \rfloor$. For example, in the first iteration, we initialize the image size $d_0 = 288$, thus when extracting 96×96 image patches, the batch size can be as big as 36. This not only increases the batch size but also allows to manual balance the data distribution. In this work, we set $\rho_c = 0.89$ as the interested anatomy only takes up 11% pixels of the whole image. As such, pathological regions are more probably to be seen in the cropped patches. Figure 6 demonstrates the details of this sampling process with two different patch sizes.

3.2 Training with Alternative Cross Validation

To make full use of the training data and avoid possible overfitting issues, we employ an alternative cross-validation strategy as part of training to predict the

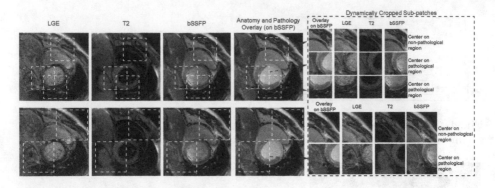

Fig. 6. Dynamic resampled image patches with varying spatial and batch sizes. The resampling sizes for images in the first and second row is 96 and 128 respectively. Smaller resampling size will bring greater batch size.

MyoPS 2020 challenge testing data. Specifically, the whole training set is split into five parts. Accordingly, the training process will be specified in five phases. In each phase, four out of the five splits are selected as the training set, while the remaining one is used as the validation set to prevent overfitting by defining the early-stopping criteria. If one phase of training is terminated, the network optimization continues on another split. The number of epochs for each training phase are 50, 40, 30, 20, and 15, while the initial learning rate are 0.0001, 0.00009, 0.00008, 0.00006, 0.00005 respectively and decayed exponentially. When all the five training phases are completed, we add a final fine-tuning phase that involves all the training data but is trained only 10 epochs with the small learning rate at 0.00004 and decayed exponentially as well. This will avoid the model to forget early trained examples.

4   Experiments

We evaluate the proposed MFU-Net on pathology segmentation using the Dice score. Experimental setup, datasets, benchmarks, and training details will be detailed in the following part.

Data: We evaluate our proposed MFU-Net on the multi-sequence CMR (MyoPS) dataset [15,16] that contains in total 25 volumes and 102 slices in the training set. For each slice, three modalities including LGE, T2, and bSSFP are provided. They are preprocessed with the Multi-variate Mixture Model [15,16], such images from the three modalities are aligned and resampled to same spatial resolution. For all the images, three anatomy masks (myocardium, left ventricle, and right ventricle) and two pathology masks (myocardial infarct and edema) are given. The testing set contains 20 volumes and 72 slices without ground-truth masks available. Both training and testing data are cropped to 288×288 to keep the region to be segmented in the sight.

Training Details: The proposed MFU-Net is optimized with fully supervised losses. The segmentation of both anatomy and pathology is trained with tversky [11] and focal [8] losses in a supervised fashion. The tversky loss is defined as $\ell_{T,j} = (\hat{y}_j^i \odot y_j^i)/[\hat{y}_j^i + y_j^i + (1-\beta)\cdot(\hat{y}_j^i - \hat{y}_j^i \odot y_j^i) + \beta\cdot(y_j^i - \hat{y}_j^i \odot y_j^i)]$ and the focal loss is $\ell_{F,j} = \sum_{H,W}[-y_j^i(1-\hat{y}_j^i)^\gamma log(\hat{y}_j^i)]$, where $\odot$ represents the element-wise multiplication and j corresponds to the involved anatomy or pathology labels. We set penalties for anatomy, infarct, and edema equal to $\lambda_{anatomy} = 1$, $\lambda_{infarct} = 3$, and $\lambda_{ana} = 5$ respectively, for each of the tversky and focal losses. Moreover, in order to achieve more stable training and quicker convergence, we initialise MFU-Net with weights from the MMSDNet [1] encoder (that also follows a U-Net architecture with dedicated encoders for each modality) when trained only with the unsupervised reconstruction loss.

Benchmarks: We evaluate the pathology segmentation performance of MFU-Net using several variants of our model. More specifically, we evaluate the effect of different design choices including the maximum fusion operator, the spatial attention module and the dynamic resampling strategy. In total we construct eight ablated models, all of which concatenate features at each encoding layer.

Table 1. Anatomy and pathology segmentation dice scores (%) of MFU-Net and relevant variants with *Residual* backbone. Myo., LV, and RV represent the myocardium, left ventricle, and right ventricle respectively. *max*, *attention*, and *resample* represent the presence of the max-fusion operator, the spatial attention module, and the dynamic resampling strategy respectively. Pathology score is calculated by averaging both the infarct and edema segmentation performance.

max	attention	resample	Myo.	LV	RV	Infarct	Edema	Avg. Pathology
✓	✓	✓	$84.3_{7.9}$	$\mathbf{87.5_{7.1}}$	$78.5_{14.2}$	$\mathbf{53.0_{20.5}}$	$28.7_{13.9}$	$\mathbf{44.9_{13.9}}$
–	✓	✓	$\mathbf{85.2_{8.1}}$	$86.8_{10.6}$	$\mathbf{78.7_{14.0}}$	$52.1_{20.4}$	$\mathbf{29.4_{12.4}}$	$42.9_{14.0}$
✓	–	✓	$84.2_{6.9}$	$86.9_{7.5}$	$76.7_{13.7}$	$46.1_{21.1}$	$28.1_{14.3}$	$41.0_{14.1}$
✓	✓	–	$84.5_{5.3}$	$87.1_{6.4}$	$74.9_{18.7}$	$49.4_{20.4}$	$\mathbf{29.4_{17.9}}$	$42.8_{15.5}$
–	–	✓	$81.1_{7.8}$	$84.2_{8.0}$	$67.2_{17.2}$	$50.2_{17.8}$	$19.3_{13.0}$	$37.5_{16.8}$
–	✓	–	$\mathbf{85.2_{4.1}}$	$86.1_{9.4}$	$75.7_{18.6}$	$52.6_{19.4}$	$28.7_{17.1}$	$43.6_{15.4}$
✓	–	–	$82.3_{7.9}$	$82.3_{8.9}$	$68.2_{16.5}$	$48.0_{25.4}$	$22.8_{15.9}$	$36.1_{18.5}$
–	–	–	$81.6_{6.5}$	$84.1_{8.0}$	$67.5_{15.5}$	$42.8_{21.7}$	$20.6_{16.6}$	$34.8_{17.7}$

4.1   Results and Discussion

We report segmentation results of MFU-Net and the ablated models in Table 1 with anatomy (myocardium, left and right ventricles) and pathology (myocardial infarct and edema) segmentation dice scores.[3] The backbone architecture used the residual connections in encoding and decoding layers [6] noted as *Residual*.

As can be seen in Table 1, the proposed maximum fusion operator and dynamic resampling achieve the best infarct segmentation, while edema segmentation performs similarly to the model without the max fusion. On average the model with all *attention, max,* and *resample* options achieves the best pathology segmentation with Dice equal to 44.9%.[4] Moreover, it can be observed that the spatial attention module improves segmentation for both infarct and edema.

In addition, the anatomy segmentation does not benefit from the proposed compositions, particularly in ventricles. The reason is two-folded. On one hand, the MyoPS 2020 challenge concentrates mainly on the pathology segmentation. As such, during training, we put more penalties on the pathology supervision (Sect. 4). It results in less focus on anatomy learning. On the other hand, because both infarct and edema is in the myocardium region, the pathology training gradient will offer an additional guide to train myocardium segmentation. On the contrary, ventricle predictions are not enjoying such an advantage.

Table 2. Anatomy and pathology segmentation comparison between *Residual, Dilation,* and *Sideconv* backbones when *max, attention,* and *resample* are all present.

	Myo.	LV	RV	Infarct	Edema	Avg. pathology
Residual	$\mathbf{84.3_{7.9}}$	$\mathbf{87.5_{7.1}}$	$\mathbf{78.5_{14.2}}$	$53.0_{20.5}$	$28.7_{13.9}$	$\mathbf{44.9_{13.9}}$
Dilation	$80.5_{4.3}$	$85.3_{6.3}$	$44.3_{33.8}$	$\mathbf{55.1_{18.7}}$	$23.1_{13.9}$	$43.7_{14.0}$
Sideconv	$76.3_{10.3}$	$65.0_{18.6}$	$40.5_{39.4}$	$52.1_{21.1}$	$\mathbf{29.7_{11.8}}$	$45.0_{16.0}$

4.2   Prediction for the Challenge Testing Dataset

Table 2 specifies the comparison with other two backbone CNN options, namely, the dilated convolutions in the bottleneck layer [14], and the side-convolution by adding 3×3, 3×1, and 1×3 convolutions in each of the convolution operations [2]. They are denoted as *Dilation* and *SideConv* respectively. It can be seen clearly that the models using dilated convolutions and side-convolutions improve on the segmentation of infarct and edema respectively, compared to our initial model using residual connections. We therefore use the *Dilation* and *Sideconv* MFU-Nets for inference of the MyoPS 2020 testing dataset. The segmentation results are presented in Table 3 and contain the Dice scores of infarct and the

---

[3] Since we do not have the ground truth of the testing data, the performance reported in Table 1 and Table 2 are obtained by five-fold cross validation across the training set. Relevant splits are following the description in Sect. 3.2. In addition, we also report the averaged pathology Dice scores of the both pathologies to assess the overall pathology segmentation performance.

[4] Although the anatomy segmentation performance decreases, we still think *SideConv* and *Dilation* are better choices since we are more caring about the pathology prediction in this research.

union of both infarct and edema. It can be seen that the *dilation* backbone with *max, attention,* and *resample* achieves better results with 58.4% dice for infarct, and 61.4% for both the infarct and the edema together.

Table 3. Pathology segmentation dice scores on the MyoPS 2020 testing data

SideConv		Dilation	
Infarct	Infarct+Edema	Infarct	Infarct+Edema
$57.0_{28.7}$	$60.3_{18.1}$	$58.4_{26.3}$	$61.4_{17.8}$

The prediction models are trained with the alternative cross validation described in Sect. 3.2. Figure 7a and Fig. 7b illustrate the training and validation dice losses respectively during model optimization. Particularly, in Fig. 7a, each loss jump corresponds to the point where the cross validation split switches and the training phase changes. Furthermore, all losses gradually decrease in each training phase, and finally converge at the final few steps.

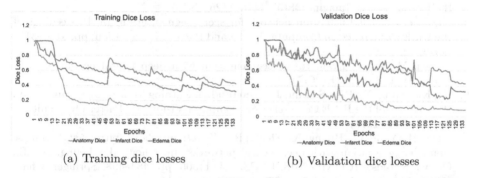

(a) Training dice losses · · · · · · · · · · (b) Validation dice losses

Fig. 7. Training and validation dice losses with the alternative cross-validation for the testing dataset. Curves in green, orange, and blue represent the anatomy, infarct, and edema dice losses.

5 Conclusions

In this paper, we proposed the Multi-Fusion U-Net, a novel architecture to segment infarct and edema from multi-modal images including LGE, T2-weighted, and bSSFP sequences. Our model uses dedicated encoders for each modality, and combines multi-modal information with feature fusion performed with the pixel-wise maximum operator at each encoding layer. These max-fused features together with the concatenated modality specific features of each encoding layer, are propagated to corresponding decoding layers of the same spatial resolution using skip connections. Additionally, a spatial attention module in the final decoding layer, as well as a novel dynamic resampling training strategy, are engaged to guide the network to focus on small pathology regions. Extensive experiments on the MyoPS 2020 challenge dataset demonstrated the effectiveness of the MFU-Net in improving cardiac pathology segmentation performance.

Acknowledgement. This work was supported by US National Institutes of Health (1R01HL136578-01). This work used resources provided by the Edinburgh Compute and Data Facility (http://www.ecdf.ed.ac.uk/). S.A. Tsaftaris acknowledges the Royal Academy of Engineering and the Research Chairs and Senior Research Fellowships scheme.

References

1. Chartsias, A., et al.: Disentangle, align and fuse for multimodal and zero-shot image segmentation. arXiv preprint arXiv:1911.04417 (2019)
2. Ding, X., Guo, Y., Ding, G., Han, J.: ACNet: strengthening the kernel skeletons for powerful cnn via asymmetric convolution blocks. In: Proceedings of the IEEE International Conference on Computer Vision, pp. 1911–1920 (2019)
3. Dolz, J., Gopinath, K., Yuan, J., Lombaert, H., Desrosiers, C., Ayed, I.B.: Hyper-DenseNet: a hyper-densely connected CNN for multi-modal image segmentation. IEEE Trans. Med. Imaging **38**(5), 1116–1126 (2018)
4. Fu, J., et al.: Dual attention network for scene segmentation. In: Proceedings of the IEEE Conference on Computer Vision and Pattern Recognition, pp. 3146–3154 (2019)
5. Havaei, M., et al.: Brain tumor segmentation with deep neural networks. Med. Image Anal. **35**, 18–31 (2017)
6. He, K., Zhang, X., Ren, S., Sun, J.: Deep residual learning for image recognition. In: Proceedings of the IEEE Conference on Computer Vision and Pattern Recognition, pp. 770–778 (2016)
7. Jiang, H., Yang, G., Huang, K., Zhang, R.: *W-Net*: one-shot arbitrary-style chinese character generation with deep neural networks. In: Cheng, L., Leung, A.C.S., Ozawa, S. (eds.) ICONIP 2018. LNCS, vol. 11305, pp. 483–493. Springer, Cham (2018). https://doi.org/10.1007/978-3-030-04221-9_43
8. Lin, T.Y., Goyal, P., Girshick, R., He, K., Dollár, P.: Focal loss for dense object detection. In: Proceedings of the IEEE International Conference on Computer Vision, pp. 2980–2988 (2017)
9. Mahmood, F., Yang, Z., Ashley, T., Durr, N.J.: Multimodal DenseNet. arXiv preprint arXiv:1811.07407 (2018)
10. Ronneberger, O., Fischer, P., Brox, T.: U-Net: convolutional networks for biomedical image segmentation. In: Navab, N., Hornegger, J., Wells, W.M., Frangi, A.F. (eds.) MICCAI 2015. LNCS, vol. 9351, pp. 234–241. Springer, Cham (2015). https://doi.org/10.1007/978-3-319-24574-4_28
11. Salehi, S.S.M., Erdogmus, D., Gholipour, A.: Tversky loss function for image segmentation using 3D fully convolutional deep networks. In: Wang, Q., Shi, Y., Suk, H.-I., Suzuki, K. (eds.) MLMI 2017. LNCS, vol. 10541, pp. 379–387. Springer, Cham (2017). https://doi.org/10.1007/978-3-319-67389-9_44
12. Takahashi, R., Matsubara, T., Uehara, K.: RICAP : random image cropping and patching data augmentation for deep CNNs. In: Asian Conference on Machine Learning, pp. 786–798 (2018)

13. Tseng, K.L., Lin, Y.L., Hsu, W., Huang, C.Y.: Joint sequence learning and cross-modality convolution for 3D biomedical segmentation. In: Proceedings of the IEEE Conference on Computer Vision and Pattern Recognition, pp. 6393–6400 (2017)
14. Vesal, S., Ravikumar, N., Maier, A.: A 2D dilated residual U-Net for multi-organ segmentation in thoracic CT . arXiv preprint arXiv:1905.07710 (2019)
15. Zhuang, X.: Multivariate mixture model for cardiac segmentation from multi-sequence MRI. In: Ourselin, S., Joskowicz, L., Sabuncu, M., Unal, G., Wells, W. (eds.) MICCAI 2016. Lecture Notes in Computer Science, vol. 9901, pp. 581–588. Springer, Cham (2016). https://doi.org/10.1007/978-3-319-46723-8_67
16. Zhuang, X.: Multivariate mixture model for myocardial segmentation combining multi-source images. IEEE Trans. Pattern Anal. Mach. Intell. **41**(12), 2933–2946 (2018)

Fully Automated Deep Learning Based Segmentation of Normal, Infarcted and Edema Regions from Multiple Cardiac MRI Sequences

Xiaoran Zhang[1] , Michelle Noga[2,3], and Kumaradevan Punithakumar[2,3]([envelope])

[1] Department of Electrical and Computer Engineering, UCLA, Los Angeles, USA
[2] Department of Radiology and Diagnostic Imaging, University of Alberta, Edmonton, Canada
punithak@ualberta.ca
[3] Servier Virtual Cardiac Centre, Mazankowski Alberta Heart Institute, Edmonton, Canada

Abstract. Myocardial characterization is essential for patients with myocardial infarction and other myocardial diseases, and the assessment is often performed using cardiac magnetic resonance (CMR) sequences. In this study, we propose a fully automated approach using deep convolutional neural networks (CNN) for cardiac pathology segmentation, including left ventricular (LV) blood pool, right ventricular blood pool, LV normal myocardium, LV myocardial edema (ME) and LV myocardial scars (MS). The input to the network consists of three CMR sequences, namely, late gadolinium enhancement (LGE), T2 and balanced steady state free precession (bSSFP). The proposed approach utilized the data provided by the MyoPS challenge hosted by MICCAI 2020 in conjunction with STACOM. The training set for the CNN model consists of images acquired from 25 cases, and the gold standard labels are provided by trained raters and validated by radiologists. The proposed approach introduces a data augmentation module, linear encoder and decoder module and a network module to increase the number of training samples and improve the prediction accuracy for LV ME and MS. The proposed approach is evaluated by the challenge organizers with a test set including 20 cases and achieves a mean dice score of 46.8% for LV MS and 55.7% for LV ME+MS.

Keywords: Cardiac magnetic resonance imaging · Deep convolutional neural network · Myocardial edema and scar · Image segmentation

1 Introduction

The imaging-based assessment of the heart using modalities such as magnetic resonance imaging (MRI) plays a central role in the diagnosis of cardiac disease, a leading cause of death worldwide. Late gadolinium-enhanced (LGE) imaging is

© Springer Nature Switzerland AG 2020
X. Zhuang and L. Li (Eds.): MyoPS 2020, LNCS 12554, pp. 82–91, 2020.
https://doi.org/10.1007/978-3-030-65651-5_8

one of the commonly used cardiac magnetic resonance (CMR) sequences to diagnose myocardial infarction [2], a common cardiac disease that may lead to heart failure. Acute injury or inflammation related to other conditions can be detected using T2-weighted CMR. However, detecting ventricular boundaries using the LGE or T2-weighted images is challenging, while this function can more easily be performed using a balanced steady state free precession (bSSFP) sequences. Often many cardiac patients are scanned using multiple CMR sequences, and utilizing the combination of these sequences will allow for obtaining robust and accurate diagnostic information [14].

The target of this study is to combine the multi-sequence CMR data to produce an accurate segmentation of cardiac regions including left ventricular (LV) blood pool (BP), right ventricular BP, LV normal myocardium (NM), LV myocardial edema (ME) and LV myocardial scars (MS) and specifically focuses on classifying myocardial pathology. Generally, the myocardium region could be divided into normal, infarcted and edematous regions. Generating accurate contour for these regions is arduous, time-consuming and thus automating the segmentation process is of great interest [10]. Zabihollahy et al. [11] proposed a semiautomatic tool for LV scar segmentation using CNNs. Li et al. [6] proposed a fully automatic tool for left atrial scar segmentation.

In this challenge, there are mainly two difficulties to produce an accurate prediction of the LV ME and MS. The first difficulty is the limited amount of training data which only consists of 25 cases. The second is the small size of the LV ME and MS regions with high intra and inter-subject variations. The inter-observer variation of manual scar segmentation is reported with a Dice score of 0.5243 ± 0.1578 [14].

In this study, we propose a fully automated approach by utilizing deep convolutional neural networks to delineate the LV BP, RV BP, LV NM, LV ME and LV MS regions from multi-sequence CMR data including bSSFP, LGE and T2. Our main contributions are the following: 1) we introduce a data augmentation module and increase the training size by 40 times using random warping and rotation; 2) we introduce a linear encoder and decoder to improve the network's training performance and utilize three different architectures including a shallow version of the standard U-net [7], Mask-RCNN [3] and U-net++[12,13] for the LV ME and LV MS block and average the predictions of the three networks followed by a binary activation with threshold 0.5. Our method is evaluated by the challenge organizers on a test set consisting of 20 cases, which contain images acquired from scanners that are not included in the training set.

2   Methodology

We introduce the pipeline shown in Fig. 1 for the LV, ME and MS segmentation. The proposed method is fully automatic and utilized no additional data other than the training set provided by the challenge organizers. The details of each module are introduced in the following sections.

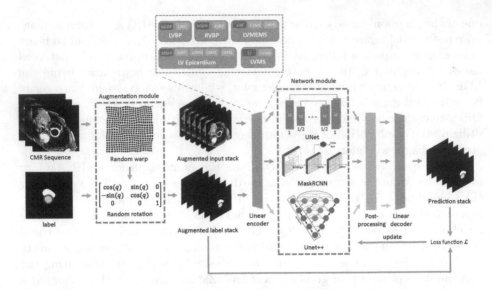

Fig. 1. Overall architecture of the proposed method in the training stage

2.1   Augmentation Module

We first extract the input data in a slice-by-slice manner and perform center cropping to obtain images of size 256×256. To improve the number of samples, we perform two data augmentation schemes including random warping and random rotation. The random warping is performed by firstly generating a $8 \times 8 \times 2$ uniformly distributed random matrix, where the last dimension indicates 2D space, with each entry in range $[-5, 5]$. We then resize the non-rigid warping matrix to the image size with dimension $256 \times 256 \times 2$ and apply the warping map using bilinear interpolation. The extracted input CMR slices and the labels are warped 20 times. After augmenting the data using random warping, we then utilize random rotation in $90°$, $180°$ and $270°$ with equal probability. The training set is then augmented with one time more data with random choice among the three options.

2.2   Preprocessing

All training and validation images are normalized using 5th and 95th percentile values, I_{05} and I_{95}, of the intensity distribution of the 2D data to obtain relatively uniform training sets. The normalized intensity value, I_n, is computed using $I_n = \dfrac{I - I_{05}}{I_{95} - I_{05}}$ where I denotes the original pixel intensity.

2.3   Linear Encoder

We introduce a linear encoder and a corresponding decoder for the augmented input stack after preprocessing. Inspired by the clinical observation in [14], we encode the augmented input and label stacks and produce five input blocks as shown in Fig. 1 instead of blindly concatenating the CMR sequences, where each block represents the data used to train a target class. The five input blocks are **LVBP block**, which uses bSSFP as the input image and LV BP as the target; **RVBP block**, which uses bSSFP as the input and RV BP as the target; **LV Epicardium block**, which uses bSSFP as the input and the linear combination of LV BP, LV NM, LV ME and LV MS as the target; **LVMEMS block**, which uses LGE as the input and the combination of LV ME and LV MS as the target; and the **LVMS block**, which uses the T2 as the input and LV MS as the target. In the testing mode, the linear encoder will only perform on the input stack. The network module will infer on the encoded input and the decoder will extract the predictions after post-processing.

2.4   Network Module

In order to improve the performance for the edema and scar prediction, we utilize three different architectures with different input blocks for each model. The results are averaged from the three networks for LV ME+MS and LV MS and followed by a binary activation with threshold 0.5. The details for each network are shown in the following.

U-Net. The U-net module utilizes a shallow version shown in Fig. 2 of the standard U-net [7] to prevent overfitting. The U-net is trained on all the five input blocks produced by the linear encoder. The loss function of the U-net is selected as the negative of dice coefficient with Adam optimizer (learning rate $= 1e-5$) and batch size $= 8$.

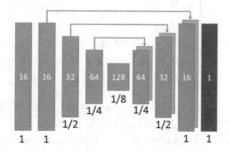

Fig. 2. Architecture of the U-net model. The blue block indicates the 3×3 convolution layer and the number indicates channel size. The blue arrow indicates the skip connection. The green block indicates the 1×1 convolution layer with sigmoid activation to produce the prediction masks. (Color figure online)

Mask-RCNN: The Mask-RCNN module [3] utilizes ResNet50 [4] as the backbone for the segmentation task and is implemented using the Matterport library [1]. The Mask-RCNN is trained on the LVMEMS and LVMS using Adam optimizer (learning rate = 0.001) and batch size = 2.

U-Net++: The U-net++ module [12,13] utilizes VGG16 [8] as the backbone. The model is trained on the LVMEMS and LVMS using the negative of dice coefficient as the loss function with Adam optimizer (learning rate = 1×10^{-5}) and batch size = 8.

2.5 Post-processing

We applied post-processing to retain only the largest connected component for the predictions of LV BP and LV Epicardium by U-net. The operation is performed in 2D space with a slice-by-slice manner. In addition, we applied an operation to remove holes that appear inside the foreground masks. As shown in Fig. 1, the post-processing is performed on the encoded predictions before the linear decoder.

2.6 Linear Decoder

The corresponding decoder performs the linear subtraction on the predicted masks of LV Epicardium and LVMEMS and is followed by a binary activation for all predicted masks in five target classes with threshold 0.5. The decoder also includes a binary constraint for the LV ME and MS predictions by calculating the myocardium mask using the predicted LV epicardium and LV BP and performing a pixelwise multiplication

$$
P_i = \begin{cases}
\sigma(\tilde{P}_0) & i = 0 \\
\sigma(\tilde{P}_1) & i = 1 \\
\sigma(\tilde{P}_2 - \tilde{P}_0 - \tilde{P}_3) & i = 2 \\
\sigma(\tilde{P}_2 - \tilde{P}_0) \odot \sigma(\tilde{P}_3 - \tilde{P}_4) & i = 3 \\
\sigma(\tilde{P}_2 - \tilde{P}_0) \odot \sigma(\tilde{P}_4) & i = 4
\end{cases}
\tag{1}
$$

where $i = 0, 1, 2, ..., 4$ denotes the index for LVBP block, RVBP block, LV Epicardium block, LVMEMS block, LVMS block respectively. P_i denotes the final prediction mask and $\tilde{P}_i$ denotes the raw prediction after post-processing for block i. $\sigma(\cdot)$ denotes the binary activation function with threshold 0.5. The notation $\odot$ denotes the pixelwise multiplication.

3 Experiments

3.1 Clinical Data

The training set consists of 25 cases of multi-sequence CMR and each refers to a patient with three sequence CMR including bSSFP, LGE and T2.

The training data is processed using the MvMM method [14,15]. The training set labels include LV BP (labelled 500), RV BP (600), LV NM (200), LV ME (1220) and LV MS (2221). The manual segmentation is performed by trained examiners and corrected by experienced radiologists. The test set consists of another 20 cases of multi-squence CMR and the ground truth is not provided.

3.2 Implementation Details

The networks are implemented using Python programming language with Keras and Tensorflow. All networks are trained with 500 epochs on NVIDIA Tesla–P100 graphical processing units with 12 GB memory. The trained neural network model with the highest validation accuracy is saved to the disk. The validation split is 0.8 for all networks with 3264 images for training and 816 images for validation after the data augmentation module. The original extracted 2D slices from the training data provided by challenge organizers contain 102 images.

3.3 Evaluation Metrics

Dice Coefficient. DC measures the overlap between two delineated regions [9]:

$$DC = \frac{2\,A\bigcap M|}{|A| + |M|} \tag{2}$$

where set A as the automatic prediction region and set M as manual segmentation ground truth.

Jaccard Index. Jaccard index measures the similarity and diversity between two delineated regions [5]:

$$J = \frac{|A\bigcap M|}{|A| + |M| - |A\bigcap M|}. \tag{3}$$

4 Results

The proposed method is evaluated over images acquired from a total of 20 cases including CMR sequences consists of bSSFP, LGE and T2. The evaluation of the proposed method on test sets are performed by the challenge organizers with dice score on LV ME+MS and LV MS. The ground truth of the test sets are not shared with the participants.

4.1 Quantitative Assessment

The agreement between the segmentation of the proposed approach with the manual ground truth is quantitatively evaluated using the dice metric and Jaccard index. To illustrate the effectiveness of the network module and the linear

Table 1. Overall performance of UNet, proposed method without the linear encoder and decoder, and the proposed method evaluated over CMR test datasets acquired from 20 cases on LV ME+MS and LV MS.

Methods	Dice metric (%)		Jaccard index (%)	
	MS	ME+MS	MS	ME+MS
UNet	36.2 ± 23.2	43.2 ± 16.0	24.5 ± 18.1	28.8 ± 13.1
Proposed†	38.5 ± 24.3	54.2 ± 17.1	26.5 ± 18.9	38.9 ± 15.0
Proposed	**46.8 ± 26.8**	**55.7 ± 18.3**	**34.2 ± 22.2**	**40.5 ± 16.3**

† indicates without the linear encoder and decoder module.

encoder and decoder, we report the performance of the UNet, proposed method without the linear encoder and decoder, and the proposed method in Table 1. The best result for the test set achieves a mean dice score of 46.8% for LV MS and 55.7% for LV ME+MS. Our proposed network module improves the overall performance of MS and ME+MS by comparing our proposed method without the linear encoder and decoder with the UNet. Our proposed linear encoder and decoder module further improves the performance especially in the MS segmentation. Figure 3 shows the performance of the three methods using box plots.

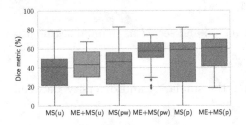 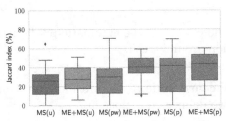

Fig. 3. The boxplots showing the performance of the proposed approach over test sets acquired from 20 cases. The evaluations were performed using Dice metric and Jaccard index. In the figure, u indicates the results by UNet; pw indicates the proposed method without linear encoder and decoder; p indicates the proposed method.

4.2 Visual Assessment

We select cases that achieve the highest and lowest dice score for visual assessment, respectively. Figure 4 shows example segmentation results where the proposed method achieved the highest agreement with the ground truth delineations. Figure 5 shows example segmentation results where the proposed method achieved the lowest agreement with the ground truth delineations.

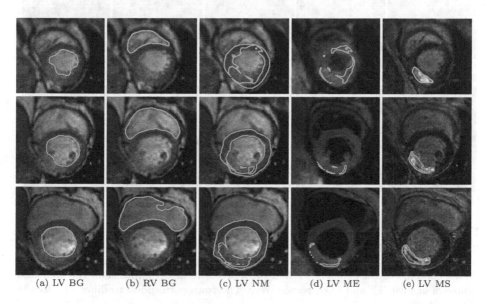

(a) LV BG   (b) RV BG   (c) LV NM   (d) LV ME   (e) LV MS

Fig. 4. Examples showing ground truth and predicted contours where the proposed method had achieved the highest dice score for LV ME+MS (75.1%) and LV MS (82.3%) with the manual delineations in the test set. The first three columns show the predicted contours against the bSSFP and the fourth and last columns show the predicted contours against T2 and LGE respectively. The rows correspond to different slices in the best case.

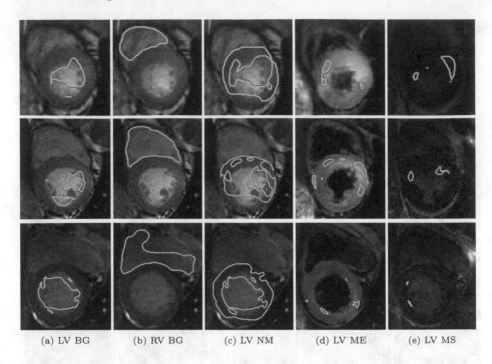

(a) LV BG   (b) RV BG   (c) LV NM   (d) LV ME   (e) LV MS

Fig. 5. Examples showing ground truth and predicted contours where the proposed method had achieved the lowest dice score for LV ME+MS (0.2%) and LV MS (30.6%) with the manual delineations in the test set. The first three columns show the predicted contours against the bSSFP and the fourth and last columns show the predicted contours against T2 and LGE respectively. The rows correspond to different slices in the worst case.

5   Conclusion

We propose a fully automated approach to segment the LV ME and LV MS from multi-sequence CMR data. We introduce an augmentation module to enhance the training set and a linear encoder and decoder along with a network module to improve the segmentation performance. The algorithm is trained using the 25 cases provided by the challenge and the evaluation is performed by the challenge organizers on another 20 cases which are not included in the training set. The proposed method yields overall mean dice metric of 46.8%, 55.7% for LV ME and LV ME+MS delineations.

Acknowledgment. The authors wish to thank the challenge organizers for providing training and test datasets as well as performing the algorithm evaluation. The authors of this paper declare that the segmentation method they implemented for participation in the MyoPS 2020 challenge has not used additional MRI datasets other than those provided by the organizers. This research was enabled in part by computing support provided by Compute Canada (www.computecanada.ca) and WestGrid.

References

1. Abdulla, W.: Mask R-CNN for object detection and instance segmentation on Keras and TensorFlow (2017). https://github.com/matterport/Mask_RCNN
2. Arai, A.E.: Magnetic resonance imaging for area at risk, myocardial infarction, and myocardial salvage. J. Cardiovasc. Pharmacol. Ther. **16**(3–4), 313–320 (2011)
3. He, K., Gkioxari, G., Dollár, P., Girshick, R.: Mask R-CNN. In: Proceedings of the IEEE International Conference on Computer Vision, pp. 2961–2969 (2017)
4. He, K., Zhang, X., Ren, S., Sun, J.: Deep residual learning for image recognition. In: Proceedings of the IEEE Conference on Computer Vision and Pattern Recognition, pp. 770–778 (2016)
5. Jaccard, P.: The distribution of the flora in the alpine zone 1. New Phytol. **11**(2), 37–50 (1912)
6. Li, L., et al.: Atrial scar quantification via multi-scale CNN in the graph-cuts framework. Med. Image Anal. **60**, 101595 (2020)
7. Ronneberger, O., Fischer, P., Brox, T.: U-net: convolutional networks for biomedical image segmentation. In: Navab, N., Hornegger, J., Wells, W., Frangi, A. (eds.) Medical Image Computing and Computer-Assisted Intervention - MICCAI 2015. LNCS, vol. 9351, pp. 234–241. Springer, Cham (2015)
8. Simonyan, K., Zisserman, A.: Very deep convolutional networks for large-scale image recognition. arXiv preprint arXiv:1409.1556 (2014)
9. Sørensen, T.J.: A method of establishing groups of equal amplitude in plant sociology based on similarity of species content and its application to analyses of the vegetation on Danish commons. I kommission hos E. Munksgaard (1948)
10. Ukwatta, E., et al.: Myocardial infarct segmentation from magnetic resonance images for personalized modeling of cardiac electrophysiology. IEEE Trans. Med. Imaging **35**(6), 1408–1419 (2016)
11. Zabihollahy, F., White, J.A., Ukwatta, E.: Convolutional neural network-based approach for segmentation of left ventricle myocardial scar from 3D late gadolinium enhancement MR images. Med. Phys. **46**(4), 1740–1751 (2019)
12. Zhou, Z., Rahman Siddiquee, M.M., Tajbakhsh, N., Liang, J.: UNet++: a nested u-net architecture for medical image segmentation. In: Stoyanov, D., et al. (eds.) DLMIA/ML-CDS -2018. LNCS, vol. 11045, pp. 3–11. Springer, Cham (2018). https://doi.org/10.1007/978-3-030-00889-5_1
13. Zhou, Z., Siddiquee, M.M.R., Tajbakhsh, N., Liang, J.: Unet++: Redesigning skip connections to exploit multiscale features in image segmentation. IEEE Trans. Med. Imaging **39**(6), 1856–1867 (2019)
14. Zhuang, X.: Multivariate mixture model for myocardial segmentation combining multi-source images. IEEE Trans. Pattern Anal. Mach. Intell. **41**(12), 2933–2946 (2019)
15. Zhuang, X.: Multivariate mixture model for cardiac segmentation from multi-sequence MRI. In: Ourselin, S., Joskowicz, L., Sabuncu, M.R., Unal, G., Wells, W. (eds.) MICCAI 2016. LNCS, vol. 9901, pp. 581–588. Springer, Cham (2016). https://doi.org/10.1007/978-3-319-46723-8_67

CMS-UNet: Cardiac Multi-task Segmentation in MRI with a U-Shaped Network

Weisheng Li$^{(\boxtimes)}$ (ORCID), Linhong Wang, and Sheng Qin

Chongqing Key Laboratory of Image Cognition, Chongqing University of Posts and Telecommunications, Chongqing, China
liws@cqupt.edu.cn

Abstract. Precise segmentation of myocardial pathology is significant for the assessment of myocardial infarction (MI). Generally, manual segmentation of myocardial pathology is burdensome and time-consuming, and the burden of disease assessment is greatly increased when considering multi-modal images. To better detect the correlations across modalities and adequately leverage the complementary information between them, we present an end-to-end architecture for automatic cardiac multi-task segmentation in magnetic resonance images (MRI) with a U-shaped network (CMS-UNet), which simultaneous segmenting left ventricular (LV) blood pool, LV myocardium, right ventricular (RV) blood pool, myocardial edema, and myocardial scars. In this work, multi-modal data are employed as the input of the network, which merely utilizes one shared encoder for extracting the feature information of different modalities respectively. Therefore, our network can automatically explore the correlations between modalities and better learn the complicated and interdependent feature representation of each modality. In decoder, we aggregated the feature information extracted from different modalities and exploited a channel reconstruction upsampling (CRU) to restore the pixel-level prediction while addressing the problem of missing more detailed information, especially for edge in bilinear upsampling. In addition, we adopted a multi-scale convolution module (MSCM) at the top of the network to capture multi-scale features, which is extremely beneficial for achieving accurate segmentation results. We evaluated our approach on the Multi-sequence CMR based Myocardial Pathology Segmentation Challenge 2020 (MyoPS 2020) dataset, and obtained the Dice 0.581 for the myocardial scars and the average Dice 0.725 for the myocardial edema and myocardial scars.

Keywords: Cardiac multi-task segmentation · MRI · Medical image segmentation

1 Introduction

Precise segmentation of myocardial pathology is significant for the assessment of myocardial infarction (MI) [1]. Cardiac magnetic resonance images (MRI) are commonly used in the diagnosis and treatment of patients who suffer from cardiovascular diseases, such as MI, in particular the balanced-Steady State Free Precession (bSSFP)

© Springer Nature Switzerland AG 2020
X. Zhuang and L. Li (Eds.): MyoPS 2020, LNCS 12554, pp. 92–101, 2020.
https://doi.org/10.1007/978-3-030-65651-5_9

cine sequence which presents distinct cardiac structure boundary, the late gadolinium enhancement (LGE) sequence can enhance the representation of infarcted myocardium, and the T2-weighted sequence which images the acute injury and ischemic regions [1, 2]. Despite advances in the medical imaging technology, most myocardial pathology segmentation tasks are still manually performed, which is burdensome, time-consuming and prone to errors. In addition, the burden of disease assessment is greatly increased when considering multi-modal image. The accurate delineation of myocardial pathology (i.e., scars) is still challenging [3–5]. Therefore, an automatic approach for cardiac segmentation is significant in clinical application.

Recently, state-of-the-art methods based on deep learning have been presented to utilize the complementary information of multi-modal data for segmentation. One way is to adopt early-fusion, which aggregate multi-modal images directly together in channel dimensions as the input of network. For example, [6] proposed an asymmetric encoder-decoder network for different acute stroke tasks such as segmentation and prediction. [7] utilizes two subnetworks to segment brain gliomas MRI. Another approach is to fuse the outputs of the separate networks to obtain the final prediction results. To name a few, a multi-path architecture [8] is proposed to exploit unique information of each modality. The work [9] adopts a multiple path 3D dense connection fully convolution neural network to capture complex combinations across modalities.

The different modal images are fused directly and fed into the network for training may hinder its expression capacity due to the difference in intensity distribution between modalities, and the design of individual encoders and even decoders for different modalities prone to cause expensive calculation cost. In this work, we propose to merely utilize one encoder for extracting the feature information of different modalities respectively. The features of different modal images are extracted with a shared encoder, and all the extracted feature information contained in different modalities are integrated together and send to the corresponding layer of decoder for feature aggregation. Since the feature representations distilled by the shared encoder are fused in the decoder, the final segmentation result of the supervisory network is equivalent to guiding the encoder to automatically explore the correlations across modalities. Hence, the network can better learn the complicated and interdependent feature representation of each modality.

In general, most state-of-the-art segmentation networks simply utilize bilinear upsampling to recover the size of the feature map after sampling to get the final segmentation results [8, 10–12]. A demerit of bilinear upsampling based on mathematical theory is its unlearnability, which leads to its limited capacity to accurately recover the pixel-wise prediction and may lose details, especially for the edge information of the object. Motivated by the work in image reconstruction [13], we present a channel reconstruction upsampling approach, termed CRU, which is prone to implement and can better restore the boundary information of feature maps.

Furthermore, we adopted a multi-scale convolution module called MSCM at the top of the network to capture different scale context information, which is extremely useful for obtaining accurate segmentation results.

2 Method

Figure 1 overviews our segmentation architecture for cardiac multi-task segmentation in MRI. The proposed left ventricular (LV) blood pool, LV myocardium, right ventricular (RV) blood pool, myocardial edema, and myocardial scars segmentation method includes four parts: shared encoder, CRU, MSCM and loss function.

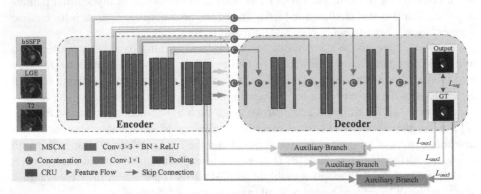

Fig. 1. The schematic diagram of Cardiac Multi-task Segmentation in MRI with a U-shaped Network (CMS-UNet). Three modalities (bSSFP, LGE and T2) of cardiac MRI are employed as network inputs respectively, we first use MSCM to obtain multi-scale contexts and a shared encoder to extract features from different modalities respectively. Then features from the corresponding layer of encoder are fused in decoder and CRU is applied to recover the size of feature maps gradually. Finally, the feature maps are fed into a convolution layer to form the final pixel-wise prediction, and the auxiliary branch help the network easier to train.

2.1 Shared Encoder

In recent years, deep learning has become the most popular and widely used method for image segmentation. As one of the state-of-the-art segmentation methods, U-Net [14] has been shown its robustness and accuracy in various medical image segmentation tasks includes brain tumors [6, 15], cardiac MRI [16–18], and pancreas [19, 20]. Hence, we improved our methods based on U-Net. Different from [8], we merely utilized one encoder to extract the feature information of each modality while three modalities are respectively employed as the input of the network, as shown in Fig. 1. To better leverage the complementary information across different modalities, and make the network have stronger data representation and discriminability, all extracted features are integrated together and fed into the corresponding layer of the decoder for feature aggregation via skip connection.

2.2 Channel Reconstruction Upsampling

Let $I \in \mathbb{R}^{C \times H \times W}$ be the input image of the network and $Y \in \mathbb{R}^{\hat{C} \times \hat{H} \times \hat{W}}$ be the ground truth which is usually one-hot encoded, where C and $\hat{C}$ denotes the number of channels of the input image and the number of classes of the ground truth, respectively. Suppose $M \in \mathbb{R}^{\tilde{C} \times \hat{H} \times \tilde{W}}$ denotes the output of the last layer of encoder, where $\tilde{H} = \hat{H}/s$, $\tilde{W} = \hat{W}/s$, and s refers to the holistic downsampling ratio, which is 16 in our network. Instead of adopting bilinear upsampling, which has limited ability to accurately recover the pixel-wise prediction, CRU reconstructs the low-resolution feature maps to high-resolution employing convolution operation thoroughly at the channel level.

The schematic diagram of CRU is shown in Fig. 2. For feature map $M \in \mathbb{R}^{\tilde{C} \times \hat{H} \times \tilde{W}}$, assuming that we expect upscale it by r times and product K feature maps, then, it is indispensable to product the feature map size of $J \times \tilde{H} \times \tilde{W}$ (N in Fig. 2), where $J = K \times r \times r$. Considering the memory and computation, instead of producing J feature maps directly, we leveraged a 3×3 convolution to obtain half of them and a cheap operation similar to [21] to get the rest. Then we concatenated all the feature maps together in terms of channel dimensions to form the output feature map $N \in \mathbb{R}^{J \times \tilde{H} \times \tilde{W}}$. Finally, we shaped N as $\mathbb{R}^{K \times \breve{H} \times \breve{W}}$ to get the final output feature map O, where $\breve{H} = \tilde{H} \times r$, $\breve{W} = \tilde{W} \times r$.

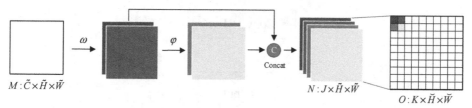

Fig. 2. The proposed channel reconstruction upsampling. ω represents the operation of convolution and φ represents the cheap operation.

CRU is a plug and play module which can be easily embedded into commonly arbitrary networks for end-to-end training. Note that to obtain better segmentation results, we suggest utilize CRU alternately in the decoder, as shown in Fig. 1.

2.3 Multi-scale Convolution Module

The lesion area in Cardiac MRI may undergo great variation in size. For instance, myocardial edema and scars size vary from patient to patient. This makes it difficult to extract information accurately with fixed size kernel. Multi-scale context information has been shown by many works to be extremely useful for obtaining better segmentation results. For example, in [22], features extracted by convolution with multiple dilation rates are concatenated together to integrate different scale feature information. [23] combines feature maps extracted by different kernel size pooling layers and concatenation operations to capture richer multi-scale global context information. Generally, these works embed multi-scale feature extraction modules at the end of the network but we put it at the top of the network because we argue that the introduction of multi-scale information at the lower layer of the network is more conducive to the final pixel-level prediction. Figure 3 depicts the overall structure of MSCM.

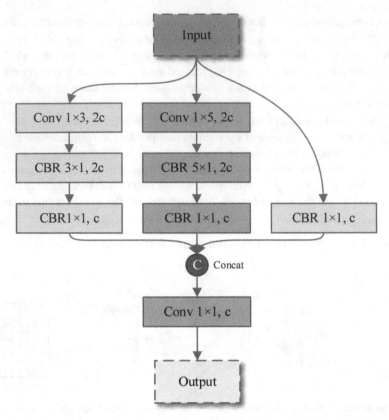

Fig. 3. Detailed architecture of multi-scale convolution module. c and CBR represents the initial channel of network and Conv + BN + ReLU, respectively.

In MSCM, 1×1, 3×3 and 5×5 kernel size convolution is adopted. Suppose the initial channels of our network is c, for 3×3 and 5×5 convolution, we expanded its output channels as $2c$ to obtain sufficient features, and then follow them with 1×1 convolution of c output channels to reduce the dimension. Finally, all feature maps are concatenated by the channel dimensions. In addition, to reduce memory consumption for MSCM, inspired by [24], we split 3×3 convolution into a combination of 1×3 and 3×1 convolution, and accordingly, 5×5 convolution is factorized to 1×5 and 5×1 convolution.

2.4 Loss Function

In our network, we used a combination of dice loss and standard binary cross-entropy (BCE) loss on predicted segmentation results p and auxiliary branch prediction results p_i:

$$L_{seg} = L_{bce}(p, \hat{y}) + L_{dice}(p, \hat{y}) \tag{1}$$

$$L_{auxi} = L_{bce}(p_i, \hat{y}) + L_{dice}(p_i, \hat{y}) \tag{2}$$

Where $i \in \{1, 2, 3\}$ and $\hat{y} \in \mathbb{R}^{\hat{C} \times \hat{H} \times \hat{W}}$ denotes the ground truth. Hence, the total auxiliary loss can be present as the following form:

$$L_{aux} = \sum_{i=1}^{N} \lambda L_{auxi} \tag{3}$$

Where λ denotes one regulable parameter which adjust the weight of the loss. Finally, the total task loss function can be expressed as:

$$L = L_{seg} + L_{aux} \tag{4}$$

3   Experimental Results

3.1   Dataset

We trained and evaluated our method on the Multi-sequence CMR based Myocardial Pathology Segmentation Challenge 2020 (MyoPS 2020) dataset, which provides 45 cases of multi-sequence CMR and each case covers three modalities of CMR (bSSFP, LGE and T2), For all cases, 25 has ground truth as a training set and 20 as a test set. Ground truth was manually annotated by experienced experts as left ventricular (LV) blood pool (labelled 500), right ventricular blood pool (600), LV normal myocardium (200), LV myocardial edema (1220), and LV myocardial scars (2221).

For the training set, we first processed each modality respectively to obtain 2D slice, and then randomly divided 80% for training and 20% for validation. For data augmentation, we adopted mirror, reverse, rotation and scale for all slices. The degrees of rotation include [30, 45, 90, 135, 180, 225, 270, 315] and the scaling factors contain [0.6, 0.7, 0.8, 0.9, 0.95, 1.1, 1.15, 1.2]. All training and validation images were normalized and ground truth was encoded with one-hot encoding. Finally, the center of each slice is cropped to 256×256.

3.2   Implementation Details

We implemented our approach on Pytorch. Specifically, we adjusted the structure of U-Net by adding an additional convolution layer to each block of encoder and appended a BN layer after each convolution layer, and set the initial channel of the network as 44 instead of 64 to reduce memory consumption and computation. This network was utilized as a baseline for comparison. We used Adam optimizer with $\beta_1 = 0.9$ and $\beta_2 = 0.999$ for parameters updating. The initial learning rate was set to 10^{-4} and weight decay of 10^{-5} was used for fine-tuning. We adopted early stopping to train the network, which finished training when the average dice of edema and scars increased less than 0.001 within 30 epochs. For the auxiliary loss L_{aux}, λ was empirically set to 0.3.

3.3 Results

We calculated the Dice coefficient between each pixel-wise prediction and ground truth to evaluate the accuracy of the segmentation results. Table 1 shows our experiment results, we first tried to train the baseline with single sequence, and the results showed that the edema and scars segmentation were best with LGE sequence and worst with bSSFP sequence, which is consistent with the fact that LGE sequence can enhance the representation of infarcted myocardium. Afterwards, using two of the three sequences for training, we observed that the segmentation results were more accurate than merely using single sequence. Finally, we noticed that employing entire sequences can further improve the segmentation results. We compared the proposed CMS-UNet with the baseline approach in the last row in Table 1, and the better segmentation performance confirms the effectiveness of our approach.

Table 1. The average Dice scores of baseline approach and CMS-UNet on the validation set. Baseline is the adjusted U-Net. LVM denotes the LV myocardium, $\sqrt{}$ and × represents use or non-use of corresponding data, respectively.

Methods	bSSFP	LGE	T2	Dice				
				LV	LVM	RV	Scars	Edema + Scars
Baseline	$\sqrt{}$	×	×	0.896	0.715	0.836	0.405	0.382
Baseline	×	$\sqrt{}$	×	0.921	0.764	0.825	0.699	0.521
Baseline	×	×	$\sqrt{}$	0.911	0.755	0.889	0.547	0.463
Baseline	$\sqrt{}$	$\sqrt{}$	×	0.930	0.781	0.892	0.641	0.533
Baseline	$\sqrt{}$	×	$\sqrt{}$	0.932	0.813	0.870	0.514	0.477
Baseline	×	$\sqrt{}$	$\sqrt{}$	0.922	0.814	0.886	0.712	0.570
Baseline	$\sqrt{}$	$\sqrt{}$	$\sqrt{}$	0.940	0.830	0.864	0.706	0.585
Proposed	$\sqrt{}$	$\sqrt{}$	$\sqrt{}$	0.931	0.828	0.879	**0.713**	**0.607**

We evaluated our method on the test set, the dice score of the proposed method achieved 0.581 for scars and 0.725 for the average of edema and scars. In addition, we visualized the segmentation results of the proposed CMS-UNet on the test set are shown in Fig. 4.

3.4 Ablation Study

To evaluate the effect of each component in our method on the final segmentation results, we conducted ablation experiments on the test set. For fairness, the baseline (i.e., U-Net with initial channels 44 and an additional convolution layer of each block in the encoder) was implemented by us, and each result was averaged over three experiments. As listed in Table 2, MSCM can achieve approximately 1% performance improvement with reference to Dice coefficient, and approximately 1.5% for CRU. In addition, as the network depth increases and the number of channels multipliers, the use of MSCM at the end of the encoder earns poor results.

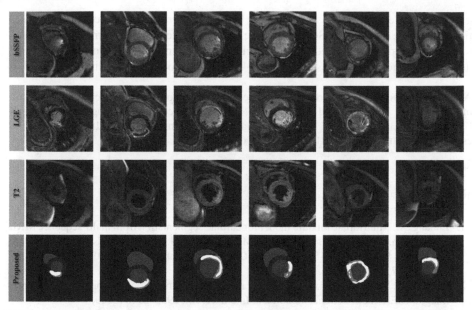

Fig. 4. Segmentation results for the proposed approach CMS-UNet on the test set. The white region represents the myocardial scars, and the light gray region near the white region represents the myocardial edema.

Table 2. Ablation study employing different modules combinations with all modalities are being as the input of the network. Baseline is the adjusted U-Net. MSCM means use multi-scale convolution module at the top of the network, MSCM* means use multi-scale convolution module at the end of the encoder and CRU means use channel reconstruction upsampling to restore the pixel-level prediction. The results are evaluated on the test set.

Method	Dice	
	Scars	Edema + Scars
Baseline	0.578 ± 0.258	0.701 ± 0.108
Baseline + MSCM	0.594 ± 0.255	0.712 ± 0.102
Baseline + MSCM*	0.578 ± 0.263	0.695 ± 0.119
Baseline + CRU	**0.605 ± 0.261**	0.717 ± 0.097
Baseline + MSCM + CRU	0.581 ± 0.268	**0.725 ± 0.110**

4 Conclusion

In this paper, we propose a multi-task segmentation network CMS-UNet for cardiac MRI segmentation. Our network can automatically detect the correlations between different modalities and learn the complicated and complementary information by utilizing a shared encoder and well-designed feature fusing manner. To obtain accurate segmentation results, we employed MSCM to capture different scale context information and CRU to recover the pixel-wise prediction. Sufficient experiments on the Myops 2020 dataset demonstrate the effectiveness of the CMS-UNet outperforms the baseline method.

Acknowledgements. This work was supported by the National Natural Science Foundation of China [Nos. 61972060 and U1713213], National Science & Technology Major Project [2016YFC1000307-3], Natural Science Foundation of Chongqing [cstc2019cxcyljrc-td0270, cstc2019jcyj-cxttX0002, cstc2019jcyj-zdxmX0011].

References

1. Zhuang, X.: Multivariate mixture model for cardiac segmentation from multi-sequence MRI. In: Ourselin, S., Joskowicz, L., Sabuncu, Mert R., Unal, G., Wells, W. (eds.) MICCAI 2016. LNCS, vol. 9901, pp. 581–588. Springer, Cham (2016). https://doi.org/10.1007/978-3-319-46723-8_67
2. Zhuang, X.: Multivariate mixture model for myocardial segmentation combining multi-source images. IEEE Trans. Pattern Anal. Mach. Intell. **41**(12), 2933–2946 (2019)
3. Zabihollahy, F., White, J.A., Ukwatta, E.: Convolutional neural network-based approach for segmentation of left ventricle myocardial scar from 3D late gadolinium enhancement MR images. Med. Phys. **46**(4), 1740–1751 (2019)
4. Li, L., Weng, X., Schnabel, J. A., Zhuang, X.: Joint left atrial segmentation and scar quantification based on a DNN with spatial encoding and shape attention. arXiv preprint arXiv: 2006.13011 (2020)
5. Li, L., et al.: Atrial scar quantification via multi-scale CNN in the graph-cuts framework. Med. Image Anal. **60**, 101595 (2020)
6. Clèrigues A., Valverde, S., Bernal, J., Freixenet, J., Oliver, A., Lladó, X.: SUNet: a deep learning architecture for acute stroke lesion segmentation and outcome prediction in multimodal mri. arXiv preprint arXiv:1810.13304 (2018)
7. Cui, S., Mao, L., Jiang, J., Liu, C., Xiong, S.: Automatic semantic segmentation of brain gliomas from MRI images using a deep cascaded neural network. J. Healthc. Eng. **2018**(1), 1–14 (2018)
8. Dolz, J., Desrosiers, C., Ben Ayed, I.: IVD-Net: intervertebral disc localization and segmentation in mri with a multi-modal UNet. In: Zheng, G., Belavy, D., Cai, Y., Li, S. (eds.) CSI 2018. LNCS, vol. 11397, pp. 130–143. Springer, Cham (2019). https://doi.org/10.1007/978-3-030-13736-6_11
9. Dolz, J., Gopinath, K., Yuan, J., Lombaert, H., Desrosiers, C., Ayed, I.B.: HyperDense-Net: a hyper-densely connected CNN for multi-modal image segmentation. IEEE Trans. Med. Imaging **38**(5), 1116–1126 (2018)
10. Zhou, Z., Siddiquee, M.M.R., Tajbakhsh, N., Liang, J.: Unet++: redesigning skip connections to exploit multiscale features in image segmentation. IEEE Trans. Med. Imaging **39**(6), 1856–1867 (2019)
11. Takikawa, T., Acuna, D., Jampani, V., Fidler, S.: Gated-SCNN: gated shape CNNs for semantic segmentation. In: Proceedings of the IEEE International Conference on Computer Vision, pp. 5228–5237 (2019)
12. Fu, J., et al.: Dual attention network for scene segmentation. In: Proceedings of the IEEE Conference on Computer Vision and Pattern Recognition, pp. 3141–3149 (2019)

13. Shi, W., et al.: Real-time single image and video super-resolution using an efficient sub-pixel convolutional neural network. In: Proceedings of the IEEE Conference on Computer Vision and Pattern Recognition, pp. 1874–1883 (2016)
14. Ronneberger, O., Fischer, P., Brox, T.: U-Net: convolutional networks for biomedical image segmentation. In: Navab, N., Hornegger, J., Wells, William M., Frangi, Alejandro F. (eds.) MICCAI 2015. LNCS, vol. 9351, pp. 234–241. Springer, Cham (2015). https://doi.org/10.1007/978-3-319-24574-4_28
15. Kamnitsas, K., et al.: Ensembles of multiple models and architectures for robust brain tumour segmentation. In: Crimi, A., Bakas, S., Kuijf, H., Menze, B., Reyes, M. (eds.) BrainLes 2017. LNCS, vol. 10670, pp. 450–462. Springer, Cham (2018). https://doi.org/10.1007/978-3-319-75238-9_38
16. Tong, Q., Ning, M., Si, W., Liao, X., Qin, J.: 3D deeply-supervised U-Net based whole heart segmentation. In: Pop, M., et al. (eds.) STACOM 2017. LNCS, vol. 10663, pp. 224–232. Springer, Cham (2018). https://doi.org/10.1007/978-3-319-75541-0_24
17. Zhou, X.Y., Yang, G.Z.: Normalization in training U-Net for 2-D biomedical semantic segmentation. IEEE Robot. Autom. Lett. 4(2), 1792–1799 (2019)
18. Chen, C., et al.: Unsupervised multi-modal style transfer for cardiac MR segmentation. In: Pop, M., Sermesant, M., Camara, O., Zhuang, X., Li, S., Young, A., Mansi, T., Suinesiaputra, A. (eds.) STACOM 2019. LNCS, vol. 12009, pp. 209–219. Springer, Cham (2020). https://doi.org/10.1007/978-3-030-39074-7_22
19. Oktay, O., et al.: Attention U-Net: learning where to look for the pancreas. arXiv preprint arXiv:1804.03999 (2018)
20. Li, F., Li, W., Shu, Y., Qin, S., Xiao, B., Zhan, Z.: Multiscale receptive field based on residual network for pancreas segmentation in CT images. Biomed. Signal Process. Control 57, 101828 (2020)
21. Han, K., Wang, Y., Tian, Q., Guo, J., Xu, C., Xu, C.: GhostNet: more features from cheap operations. In: Proceedings of the IEEE Conference on Computer Vision and Pattern Recognition, pp. 1580–1589 (2020)
22. Chen, L.C., Papandreou, G., Schroff, F., Adam, H.: Rethinking atrous convolution for semantic image segmentation. arXiv preprint arXiv:1706.05587 (2017)
23. Zhao, H., Shi, J., Qi, X., Wang, X., Jia, J.: Pyramid scene parsing network. In: Proceedings of the IEEE Conference on Computer Vision and Pattern Recognition, pp. 2881–2890 (2017)
24. Szegedy, C., Vanhoucke, V., Ioffe, S., Shlens, J., Wojna, Z.: Rethinking the inception architecture for computer vision. In: Proceedings of the IEEE Conference on Computer Vision and Pattern Recognition, pp. 2818–2826 (2016)

Automatic Myocardial Scar Segmentation from Multi-sequence Cardiac MRI Using Fully Convolutional Densenet with Inception and Squeeze-Excitation Module

Tewodros Weldebirhan Arega[1,2,3]([✉]) [iD] and Stéphanie Bricq[1] [iD]

[1] ImViA Laboratory, Université Bourgogne Franche-Comté, Dijon, France
tewdrosw@gmail.com
[2] University of Girona, Girona, Spain
[3] University of Cassino and Southern Lazio, Cassino, Italy

Abstract. Automatic and accurate myocardial scar segmentation from multiple-sequence cardiac MRI is essential for the diagnosis and prognosis of patients with myocardial infarction. However, this is difficult due to motion artifact, low contrast between scar and blood pool in late gadolinium enhancement (LGE) MRI, and poor contrast between edema and healthy myocardium in T2 cardiac MRI. In this paper, we proposed a fully-automatic scar segmentation method using a cascaded segmentation network of three Fully Convolutional Densenet (FC-Densenet) with Inception and Squeeze-Excitation module. It is called Cascaded FCDISE. The first FCDISE is used to extract the region of interest and the second FCDISE to segment myocardium and the last one to segment scar from the pre-segmented myocardial region. In the proposed segmentation network, the inception module is incorporated at the beginning of the network to extract multi-scale features from the input image, whereas the squeeze-excitation blocks are placed in the skip connections of the network to transfer recalibrated feature maps from the encoder to the decoder. To encourage higher order similarities between predicted image and ground truth, we adopted a dual loss function composed of logarithmic Dice loss and region mutual information (RMI) loss. Our method is evaluated on the Multi-sequence CMR based Myocardial Pathology Segmentation challenge (MyoPS 2020) dataset. On the test set, our fully-automatic approach achieved an average Dice score of 0.565 for scar and 0.664 for scar+edema. This is higher than the inter-observer variation of manual scar segmentation. The proposed method outperformed similar methods and showed that adding the two modules to FC-Densenet improves the segmentation result with little computational overhead.

Keywords: Multi-sequence cardiac MRI · Myocardial scar · Segmentation · Deep learning · Fully-automatic

© Springer Nature Switzerland AG 2020
X. Zhuang and L. Li (Eds.): MyoPS 2020, LNCS 12554, pp. 102–117, 2020.
https://doi.org/10.1007/978-3-030-65651-5_10

1 Introduction

Cardiovascular diseases (CVDs) are the number one cause of death globally [12]. Myocardial infarction (MI), commonly known as a heart attack, is the irreversible death of heart muscle (myocardium) due to lack of oxygen supply (ischemia) [2].

Cardiac magnetic resonance (CMR) is a set of magnetic resonance imaging (MRI) often used to provide anatomical and functional information of the heart. Late gadolinium enhancement (LGE) is one type of CMR which is a gold standard for visualization and quantification of myocardial infarction. T2-weighted CMR is mostly used to visualize myocardial edema whereas balanced Steady State Free Precession (bSSFP) cine sequence has clear myocardial boundaries. These sequences bring complimentary information to each other.

Myocardial scar is often segmented manually in a clinical routine. However, manual segmentation is very exhausting and suffers from intra- and inter-observer variability. This problem can be addressed by developing an automatic segmentation method. Having said that, automatic segmentation also comes with its own challenges. Heterogeneous intensity distributions of the images, large shape and size variation of the heart, motion artifact, low contrast between scar and blood pool in LGE as well as low contrast between edema and healthy myocardium in T2 make developing automatic segmentation methods difficult.

Most scar segmentation studies can be categorized into two main groups: non-deep learning based and deep learning based methods. The non-deep learning based approaches are mainly focused on thresholding and clustering. The threshold based approaches exploit the enhanced intensity of the infarcted myocardium compared with the healthy myocardium. A thresholding method called Full Width at Half Maximum (FHFW) defines a threshold value as the half value of the infarcted myocardium's maximum intensity [1]. Another method [9] defines the threshold as an intensity value n standard deviations higher than the mean intensity of the healthy myocardium (nSD); where n can be between 2 and 6. Both methods are simple, however, they require manual interaction of a user to determine region of interest that defines the threshold values. Other groups [3,13] used clustering based approach to segment the scar.

Recently, few studies have been proposed to segment scar using semi-automatic and fully-automatic deep learning methods. Zabihollahy *et al.* (2018) used manual segmentation for myocardium and then 2D Fully Convolutional Network (FCN) to segment scar from the myocardium [18]. Moccia *et al.* (2019) proposed semi-automatic and fully-automatic scar segmentation method [10]. Their semi-automatic approach, which manually segments the myocardial region, performed better than the one that uses automatic approach due to the mediocre segmentation performance of the network on myocardium. Another fully-automatic approach [15] uses a 2D U-net based myocardium segmentation followed by a top-hat transforms based coarse scar segmentation and finally a voxel classification of healthy and infarcted myocardium. However, using morphological operation to segment a scar can be unreliable particularly when the images have heterogeneous intensity distribution and motion artifact.

In this paper, we proposed a fully-automatic scar segmentation method using a cascaded Fully Convolutional-Densenet (FC-Densenet) [7] with Inception [16] and Squeeze-Excitation (SE) modules [5]. The input to our method was a multimodal image which consists of LGE, T2 and bSSFP CMR sequences.

Our work has the following main contributions: 1) We proposed three cascaded segmentation networks that extract the region of interest then segment myocardium and finally segment scar from pre-segmented myocardial region. This resulted in higher Dice score and lower false positives compared to the one that uses 2 cascaded networks. 2) We showed that incorporating SE blocks and inception module to FC-Densenet improves the segmentation performance with little computational overhead. SE blocks are incorporated in the skip connections of the network to transfer a recalibrated feature maps from encoder to decoder and inception module is added at the beginning of the network to extract multi-scale features from the input image. 3) We proposed a novel loss function that combines the conventional logarithmic Dice loss with region mutual information (RMI) loss [19]. This objective function can be useful to segment small structures and pixels with weak visual evidence such as myocardial scar and edema. 4) Our fully-automatic approach showed a promising result on Multi-sequence CMR based Myocardial Pathology Segmentation (MyoPS 2020) challenge dataset by achieving a higher Dice score for scar than the inter-observer variation of manual scar segmentation.

2 Materials

The dataset used in this paper was Multi-sequence CMR based Myocardial Pathology Segmentation Challenge (MyoPS 2020)[1]. It is part of Statistical Atlases and Computational Modeling of the Heart (STACOM) 2020 workshop and MICCAI 2020. The dataset consists of three sequence CMR of 45 subjects diagnosed with myocardial infarction. From the 45 subjects, 25 of them are used for training and the rest for testing. The sequences are LGE CMR, T2-weighted CMR and bSSFP cine sequence. LGE CMR is a T1-weighted, inversion-recovery, gradient-echo sequence. The bSSFP CMR is a balanced steady-state, free precession cine sequence and T2 CMR is a T2-weighted, black blood Spectral Presaturation Attenuated Inversion-Recovery (SPAIR) sequence. The three sequences were breath-hold and scanned at end-diastolic phase. They were also acquired in the ventricular short-axis views [20,21]. The typical parameters of the three sequences are summarized in Table 1.

The three CMR sequences were registered into a common space and similar spatial resolution with a mean of 0.75×0.75 mm using multivariate mixture model (MvMM) method [21]. All images have annotation for right ventricle, left ventricle, myocardium, scar and edema. In this paper, we focused on segmentation of all except right ventricle.

---

[1] http://www.sdspeople.fudan.edu.cn/zhuangxiahai/0/myops20/.

As a pre-processing step, the intensity of every patient image is normalized to have zero-mean and unit-variance. The dataset is already registered, as mentioned before. However, there are slight variations of spatial resolution among the patients (0.72–0.76 mm). To account for this, all patients were re-sliced to have the same spatial resolution of 1.0×1.0 mm. The z spacing of the voxel spacing is not changed.

Table 1. MRI parameter setting for bSSFP, LGE and T2 CMR sequences

Parameter	bSSFP	LGE	T2
TR/TE	2.7/1.4 ms	3.6/1.8 ms	2000/90 ms
Slice thickness	8–13 mm	5 mm	12–20 mm
In-plane resolution	1.25×1.25 mm	0.75×0.75 mm	1.35×1.35 mm

3   Methods

3.1   Proposed Pipeline

The proposed pipeline consists of data pre-processing and deep learning based region of interest extraction, myocardium and scar segmentation (Fig. 1). In our approach, a cascaded segmentation network consisting of three FC-Densenet with Inception and Squeeze-Excitation module (Cascaded FCDISE) were used to extract the region of interest and then segment myocardium and finally segment scar from the pre-segmented myocardial region. The segmentation network architecture used for the three tasks are almost the same. The only differences are the number of pooling/upsampling layers and their weights as they are trained independently. The segmentation network is based on 2D convolution operations.

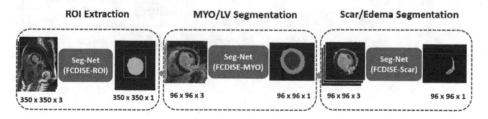

Fig. 1. Proposed pipeline. **FCDISE-ROI**: segmentation network used for ROI extraction, **FCDISE-MYO**: segmentation network used for myocardium segmentation, **FCDISE-Scar**: segmentation network used for scar segmentation

3.2 Network Architecture

The proposed method is based on FC-Densenet [7]. To enhance FC-Densenet's performance, we incorporated two important modules: SE blocks and inception module. We named the proposed segmentation network FCDISE.

FC-Densenet is an extension of Densenet [6] that deals with semantic segmentation task. Densenets are good fit for semantic segmentation because they have skip connections and multi-scale supervision by design. However, directly extending Densenet as Fully Convolutional Network (FCN) will lead to feature map explosion in the decoder part. To mitigate this problem, only the feature maps created by the preceding dense block are upsampled. Like FCN, skip connections are used to transfer the higher resolution information from encoder to decoder [7].

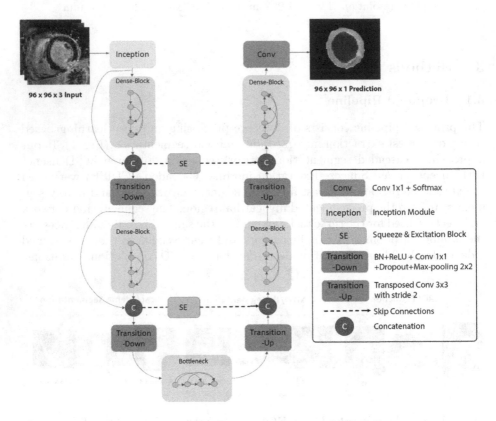

Fig. 2. Proposed network architecture (FCDISE)

Similar to FC-Densenet, our network architecture consists of downsampling path, upsampling path and skip connections. The downsampling path is composed of dense blocks and transition down layers as shown in Fig. 2. The upsampling path also has dense blocks and transition up layers. In the dense block, each layer receives feature maps from all preceding layers and forwards its feature map to all subsequent layers as shown in Fig. 3(a). Each dense block layers are made up of Batch Normalization, rectified linear unit (ReLU) activation function, 3×3 convolution and drop out with probability 0.2 (Fig. 3(b)). Transition down is composed of Batch Normalization, ReLU activation function, 1×1 convolution, drop out with probability 0.2 and 2×2 max-pooling layer with stride 2 to down-sample the feature maps into latent space. From the latent space, transition up recovers the input spatial resolution by upsampling the feature maps using 3×3 transposed convolution with stride 2. Skip connections are used to concatenate the feature maps from downsampling path to the corresponding feature maps in the upsampling path.

As shown in Fig. 2, SE blocks are incorporated in the skip connections of our network architecture. SE block is used to model channel relationships and it is regarded as self-attention on the channels. SE block consists of global average pooling and fully connected (FC) layers [5]. The SE block in our network receives the feature maps from encoder and then recalibrates the feature maps before concatenating them to the corresponding feature maps in the decoder.

The second module integrated to FC-densenet is inception module. Inspired by [8], the inception module is incorporated at the beginning of the network. The inception module in our network is a bit modified from the naive inception module [16] as it contains only three kernels (3×3, 5×5 and 7×7 kernels) and their output is summed instead of concatenated because summation yielded better results (Fig. 3(c)). The reason we used inception module as first layer of the network is to extract multi-scale features simultaneously from the input image using different sized kernels and to send the aggregated features to the next stage. These kernels can help to capture relevant features in different sizes

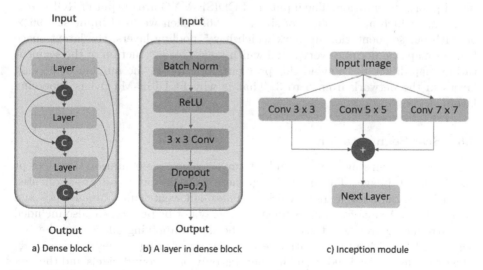

Fig. 3. Diagram of (a) dense block, (b) a layer in dense block and (c) an inception module used in our model.

of scar and edema. This can be beneficial because heart size varies from one patient to another and even in one patient there is variation of size from apex to base.

The segmentation network used to detect ROI is called FCDISE-ROI. It has 5 pooling layers. For myocardium and scar segmentation, we employed FCDISE-MYO and FCDISE-Scar respectively. Both of them have 3 pooling layers.

3.3   Region of Interest (ROI) Detection

The first stage in the proposed pipeline is ROI extraction. In the full size cardiac MR images, the heart covers very small part of the image. Due to this, it is necessary to extract a region of interest around the ventricles before proceeding to the next stages in the pipeline. Our ROI extraction method is done by first segmenting the epicardial region from the full-size cardiac MR using FCDISE-ROI. Then the center of the segmented epicardial region is calculated. Finally, we applied center cropping from the computed center of epicardial region with a patch size of 96 × 96. This particular size is chosen after taking into consideration the largest diameter of epicardium from the training set images.

This method places the ventricles in the center of the cropped region. This has three advantages for the next stages in the pipeline. It reduces the false positives and alleviates the class imbalance between the background and the ventricles/scar classes. Furthermore, it decreases the computation time as the size of input images are decreasing.

3.4   Myocardium and Left Ventricle Segmentation

The second stage in the proposed pipeline is myocardium and left ventricular blood pool segmentation. The inputs to FCDISE-MYO are output of ROI detection stage which are 2D slices of size 96 × 96. When we used input size 96 × 96 with our segmentation network which has 5 pooling layers, the latent space feature map size becomes very small which makes reconstruction of the segmentation map difficult. To avoid this problem, we reduced the number of pooling layers in the network from 5 to 3. That is why FCDISE-MYO has 3 pooling layers.

3.5   Scar Segmentation

Scar segmentation stage is very similar to myocardium segmentation stage except for the input image. The input image here contains only the pre-segmented epicardial region, the region which includes left ventricular blood pool and myocardium. As myocardium segmentation may not be perfect, we also included the surrounding area near the epicardium border by applying dilation on the pre-segmented epicardial region with a rectangular structuring element of size 5 × 5. The input image size is 96 × 96 but contains only background pixels and the pre-segmented region. The segmentation network used in this stage is FCDISE-Scar, which is similar to the previous stage's segmentation network.

As a post-processing step, we applied 2D connected component analysis and morphological operations like dilation and erosion to the segmented image to further improve the segmentation result and reduce outliers.

3.6   Loss Function

As an objective function, we proposed a dual loss function which is a weighted combination of logarithmic Dice loss [17] and region mutual information (RMI) loss [19].

Logarithmic Dice loss (log Dice loss) is known for its robust performance on small structures [17]. Compared to linear Dice loss, it focuses more on less accurate classes. Log Dice loss is computed as the mean value of the natural logarithm of the Dice coefficient as stated in Eq. 1. It also introduces an exponent γ that controls the non-linearity of the loss function. When $\gamma > 1$, the log Dice loss focuses even more on the less accurate classes. If the non-linearity is $0 < \gamma < 1$, the loss works better because it supports improvement at both low and high accuracy. To improve the segmentation of small structures like scar and edema, we chose a logarithmic Dice loss.

The second loss function used is region mutual information loss. Unlike pixel-wise loss, RMI loss takes into account the dependencies among the pixels. Each pixel in an image is represented by the pixel itself and its neighbouring pixels. In other words, the pixel will be represented by multi-dimensional point and the image will be a multi-dimensional distribution of these points. Maximizing the mutual information between the multi-dimensional distributions of the ground truth and predicted image will result in high order consistency between these two images. This loss function captures the structural differences between the shapes of predictions and ground truth. It is also helpful in identifying pixel whose visual evidence is weak or when the pixel belongs to objects with small spatial structures [19]. This makes it ideal for myocardium and scar segmentation.

From Eq. 2, Y is multi-dimensional distribution of the ground truth and P is multi-dimensional distribution of the predicted image. $\Sigma_{Y|P}$ is the posterior covariance matrix of Y given P and $det()$ is determinant of the matrix. $I(Y; P)$ is a lower bound of the mutual information. Then the total RMI loss is computed as a combination of the pixel-wise cross entropy loss (L_{CE}) and lower bound MI as stated in Eq. 3. In this equation, B and C represent mini-batch size and number of classes respectively.

To take advantage of both log Dice loss and RMI loss, we used a weighted combination of these two losses as our objective function as stated in Eq. 4, where λ_{Dice} and λ_{RMI} are the weighting factors for log Dice loss (L_{Dice}) and RMI loss (L_{RMI}) respectively.

$$L_{Dice} = E[(-ln(Dice_i)^\gamma] \tag{1}$$

$$I(Y; P) = -\frac{1}{2}log((2\pi e)^d det(\Sigma_{Y|P})) \tag{2}$$

$$L_{RMI} = L_{CE} + \frac{1}{B} \sum_{b=1}^{B} \sum_{c=1}^{C} (-I(Y;P)) \tag{3}$$

$$L_{Total} = \lambda_{Dice} L_{Dice} + \lambda_{RMI} L_{RMI} \tag{4}$$

3.7   Training

The three segmentation networks in the pipeline are trained independently. The weights are initialized using *He normal* initialization method [4]. The optimization of the weights are done using Adam optimizer with learning rate of 0.001. The mini-batch size was 16. The model was trained for 80 epochs. We set a weighting factor of 0.8 for log Dice loss and 0.2 for RMI loss as they provided the best results. For log Dice loss, a non-linearity of 0.3 was used. The frameworks used to implement the model and the code are PyTorch and Python.

In order to avoid over-fitting, we have adopted three techniques: dropout, early stopping and weight regularization. In our experiments, the patience for the early stopping was 10 epochs and L2 weight regularization was used with regularization term (lambda) set to 1×10^{-8}. Furthermore, we used a dropout with probability of 0.2.

4   Results and Discussion

To evaluate the segmentation results, we used Dice coefficient and Hausdorff distance (HD). Dice coefficient measures the similarity of two images. It is calculated as the size of the overlap between segmented image and ground truth divided by the total size of the two images. This measures the overall quality of a segmentation. This metrics is used to evaluate both scar and myocardium segmentation results. Hausdorff distance is the greatest of all distances from a point in one set to the closest point in the other set. This metrics focuses on outliers. Hausdorff distance metric (2D) is used to evaluate myocardium segmentation result. Calculating Hausdorff distance for scar and edema can be difficult because they are dispersed regions.

To evaluate our models, we employed a five fold cross-validation as well as train-validation-test evaluation methods. For the latter method, from a total of 25 subjects, 17 were used for training, 3 for validation and 5 for test.

4.1   Myocardium and Left Ventricle

The proposed method yielded a Dice score of 0.872 and Hausdorff distance (2D) of 3.392 mm on myocardium (MYO) segmentation and a Dice score of 0.921 and Hausdorff distance (2D) of 2.577 mm on left ventricle (LV) segmentation.

The inter-observer variation of manual segmentation of MYO were Dice scores of 0.757, 0.824 and 0.812 for LGE, T2 and bSSFP respectively. Comparing to our model's performance on each CMR separately, our method yielded Dice scores of 0.771, 0.798 and 0.854 for MYO using LGE, T2 and bSSFP sequences

respectively. This result was on average better than the inter-observer variation. Besides, the Dice score of MYO increased to 0.872 when we combined the
three modalities as an input to our method. From this, we can say that combining multiple CMR modalities improved the segmentation accuracy of the heart
structures.

To evaluate the effect ROI in our pipeline, we compared the results with and
without ROI. When we directly segment heart from the full-sized cardiac MR,
our method yielded Dice scores of 0.905 and 0.853 for LV and MYO respectively.
However, when we employed ROI, our method achieved an improved Dice score
of 0.921 for LV and 0.872 for MYO. Moreover, the obtained Hausdorff distance
was on average 0.22 mm lower than the one that did not use ROI. This shows
how extracting ROI can improve the result by reducing the false positives and
mitigating the class imbalance problem between the background and ventricle
classes.

Table 2 quantitatively compares the proposed loss with the conventional
loss functions such as cross-entropy loss, Dice loss, logarithmic Dice loss. The
proposed loss outperformed the other loss functions by achieving the highest Dice
score in both LV and MYO. To better investigate the qualitative performance
of the loss functions, we selected a typically challenging image which has scar
tissue, as depicted in Fig. 4. The proposed loss produced robust segmentation
result. The other loss functions particularly failed because they segment the scar
as blood pool instead of MYO (middle slice in Fig. 4). That is when the addition
of RMI loss becomes very handy. Because RMI loss takes into account the pixel
dependencies unlike the pixel-wise losses. This helped the model to achieve high
order consistency between the prediction and ground truth.

Table 2. Quantitative comparison of loss functions using Dice score

Loss function	LV (Dice)	MYO (Dice)
Cross-entropy	0.905 ± 0.067	0.849 ± 0.086
Dice loss	0.903 ± 0.073	0.858 ± 0.067
Log dice loss	0.909 ± 0.056	0.865 ± 0.054
Proposed loss	$\mathbf{0.921 \pm 0.041}$	$\mathbf{0.872 \pm 0.041}$

4.2 Scar

The performance of the proposed method in scar, edema and scar+edema segmentation is presented in Table 3. Note that *Scar+Edema* considers scar and
edema as one class. Having one class can be helpful to evaluate the model's
performance on detecting the infarcted myocardium in general instead of divid-
ing the infarcted region into scar and edema. Our method performed well on
infarcted myocardium (scar+edema) segmentation. However, its performance
decreased a little bit when separately segmenting scar and edema.

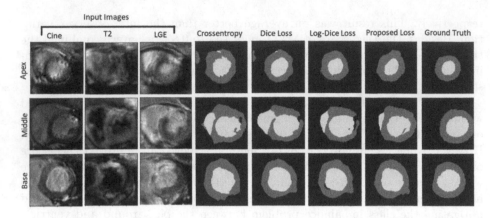

Fig. 4. Qualitative comparison of loss functions on a typically challenging image. Note that the results are before post-processing. Myocardium (green) and left ventricle (yellow). (Color figure online)

Similar to myocardium segmentation, we studied the effect of using single modal CMR and multi-modal CMR as shown in Fig. 5. Comparing the three modalities, using only LGE CMR achieved the best Dice score for scar (0.603) whereas using only T2 CMR yielded the best result for scar+edema (0.644). The bSSFP sequence's segmentation performance on both scar and edema was inferior compared to the other two CMR sequences. This can be due to the fact that bSSFP CMR has less information about scar and edema. When we combined the three modalities, the Dice score of scar slightly increased to 0.604 while that of scar+edema significantly increased to 0.687. This showed that the three CMR sequences have complementary information about scar and edema.

Table 3. Scar, edema and scar+edema segmentation result of the proposed method

Metrics	Scar	Edema	Scar+edema
Dice	0.604 ± 0.167	0.488 ± 0.172	0.687 ± 0.072
Specificity	0.977 ± 0.092	0.967 ± 0.112	0.962 ± 0.081
Sensitivity	0.627 ± 0.128	0.457 ± 0.125	0.739 ± 0.094
Accuracy	0.959 ± 0.093	0.946 ± 0.113	0.941 ± 0.098

Comparing the performance of the loss functions on the segmentation of scar and edema, the proposed loss outperformed the conventional loss functions. Cross-entropy loss yielded Dice scores of 0.527 for scar and 0.567 for scar+edema whereas Dice loss achieved Dice scores of 0.543 for scar and 0.575 for scar+edema. Log Dice loss, compared to the first two losses, provided better result for both scar (0.588) and scar+edema (0.606). When we combined RMI loss with log Dice loss, the segmentation result of scar increased a little bit to 0.604 while the

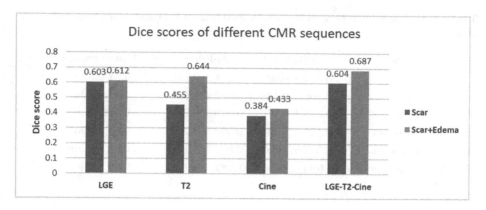

Fig. 5. Comparison of different cardiac MR sequences performance on scar and scar+edema segmentation. Where LGE is late gadolinium enhancement cardiac MR and T2 is T2-weighted cardiac MR. Cine is bSSFP cine sequence and LGE-T2-Cine is multi-modal image consisting of LGE, T2 and bSSFP sequences.

improvement for scar+edema was substantial as it enhanced the Dice score from 0.606 to 0.687. It is observable that the addition of RMI loss helped to improve the results particularly that of edema. The proposed loss function's robust segmentation performance on scar/edema and myocardium verified the benefit of combining log dice loss with a loss function that considers the dependencies among the pixels.

Ablation Study. To evaluate the effect of addition of inception and SE module to FC-Densenet, we have compared the proposed method with FC-Densenet and FC-Densenet with only SE module (FCDensenet_SE). As presented in Table 4, the baseline model (FC-Densenet) achieved comparable result in scar but failed in Edema. Adding SE blocks to the baseline has substantially improved the segmentation accuracy (Dice score) for scar+edema by nearly 10%. While the proposed method, which adds both SE block and inception module to the baseline, improved the Dice value for scar+edema achieving a 14% increase compared to the baseline. The improvement is also demonstrated in the qualitative result as can be seen from Fig. 6. It can be observed that the proposed method has comparatively better performance at detecting different sized scars. This showed the benefit of the extracted multi-scale features from the input image and confirmed the advantage of the incorporated SE block. Our method achieved this enhancement with minimal computational overhead.

Comparison with Alternative Methods. We compared our proposed method with three other methods which employed the same pipeline that is a cascaded three networks. The segmentation networks used in the place of FCDISE are Unet [14], Attention-Unet [11] and Res-Unet. Unet is one of the most commonly used segmentation networks in medical images. Attention-Unet

Table 4. Dice score comparison of various methods for scar and scar+edema segmentation.

Methods	Scar (Dice)	Scar+Edema (Dice)	No of params
Unet	0.577 ± 0.095	0.558 ± 0.131	0.84 million
Attention-Unet	0.566 ± 0.144	0.610 ± 0.118	2.6 million
Res-Unet	0.535 ± 0.176	0.560 ± 0.284	6.7 million
FCDensenet	0.579 ± 0.148	0.540 ± 0.229	0.65 million
FCDensenet_SE	0.584 ± 0.181	0.640 ± 0.134	0.68 million
Proposed method	$\mathbf{0.604 \pm 0.167}$	$\mathbf{0.687 \pm 0.072}$	0.69 million

is a standard Unet with attention gate which recalibrate feature maps spatially. Res-Unet is also a Unet with residual encoder and decoder.

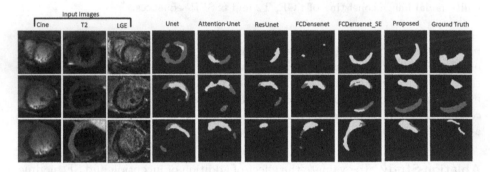

Fig. 6. Qualitative comparison of different models on scar (yellow) and edema (green) segmentation.

The comparison was both qualitatively and quantitatively, as shown in Fig. 6 and Table 4 respectively. Unet had good result on scar but its performance decreased on edema. While Res-Unet did not perform well on both scar and edema. This is because it overfitted on this small dataset (25 cases). Both Attention-Unet and the proposed method which use attention on feature maps achieved better result on scar+edema than the ones that do not use. This demonstrated the benefits of recalibrating feature maps spatially or channel-wise on helping the model to increase its focus on scar and edema. However, when Attention-Unet is compared to the proposed method, our method achieved more accurate segmentation performance in both scar and scar+edema. Besides, the proposed method was robust at detecting scar at different heart positions and had less false positive cluster of scar compared to the other methods.

As shown in Table 4, we also compared the number of trainable parameters. The ones that use dense blocks have the lowest number of parameters because Densenet encourages feature reuse which substantially reduces the number of parameters. Besides, our method is ideal on tasks with smaller training set sizes like MyoPS 2020 because the dense connections in the network have a regularizing effect which reduces overfitting.

5   Conclusion

In this paper, we proposed a deep learning based fully automatic myocardial scar segmentation method from multi-sequence cardiac MR images. Our method employs three cascaded segmentation networks to first extract ROI then segment myocardium and finally use the pre-segmented myocardium to segment scar and edema. Each segmentation network used FC-Densenet with Inception and Squeeze-Excitation module (FCDISE). The SE blocks are incorporated in the skip connections and the inception module is added in the initial layer of the network to concatenate different field of views of image features. We demonstrated that adding these two modules to FC-Densenet substantially improves the segmentation result with little computational overhead. Compared to other similar networks, our method is better at locating different size scar and edema, and performs well on small training set. Furthermore, we showed that region mutual information loss combined with logarithmic Dice loss achieves high order consistency between the prediction and ground truth. It can also be of great interest for segmentation of medical organs whose pixels have weak visual evidence.

Despite having a very challenging dataset, our approach yielded very good result on the test set achieving an average Dice score of 0.565 for scar and 0.664 for scar+edema which is higher than the inter-observer variation of scar segmentation 0.524 (Dice score of scar). Note that our proposed method's performance on few cases (3 out of 20 test cases) was poor because these cases were exceptionally challenging. Future work will aim in using multi-planar network that will include sagittal, coronal and axial views to further improve the segmentation result.

Acknowledgements. T.W. Arega received an Erasmus+ scholarship from the Erasmus Mundus Joint Master Degree in Medical Imaging and Applications (MAIA), a program funded by the Erasmus+ program of the European Union. This work was also supported by the French National Research Agency (ANR), with reference ANR-19-CE45-0001-01-ACCECIT.

References

1. Amado, L.C., et al.: Accurate and objective infarct sizing by contrast-enhanced magnetic resonance imaging in a canine myocardial infarction model. J. Am. Coll. Cardiol. **44**(12), 2383–2389 (2004)
2. Belleza, M.: Myocardial infarction: Nursing management and study guide (2017). https://nurseslabs.com/myocardial-infarction/. Accessed 10 June 2020
3. Detsky, J.S., Paul, G., Dick, A.J., Wright, G.A.: Reproducible classification of infarct heterogeneity using fuzzy clustering on multicontrast delayed enhancement magnetic resonance images. IEEE Trans. Med. Imaging **28**(10), 1606–1614 (2009)

4. He, K., Zhang, X., Ren, S., Sun, J.: Delving deep into rectifiers: surpassing human-level performance on ImageNet classification. In: Proceedings of the IEEE International Conference on Computer Vision, pp. 1026–1034 (2015)
5. Hu, J., Shen, L., Sun, G.: Squeeze-and-excitation networks. In: Proceedings of the IEEE Conference on Computer Vision and Pattern Recognition, pp. 7132–7141 (2018)
6. Huang, G., Liu, Z., Van Der Maaten, L., Weinberger, K.Q.: Densely connected convolutional networks. In: Proceedings of the IEEE Conference on Computer Vision and Pattern Recognition, pp. 4700–4708 (2017)
7. Jégou, S., Drozdzal, M., Vazquez, D., Romero, A., Bengio, Y.: The one hundred layers Tiramisu: fully convolutional DenseNets for semantic segmentation. In: Proceedings of the IEEE Conference on Computer Vision and Pattern Recognition Workshops, pp. 11–19 (2017)
8. Khened, M., Kollerathu, V.A., Krishnamurthi, G.: Fully convolutional multi-scale residual DenseNets for cardiac segmentation and automated cardiac diagnosis using ensemble of classifiers. Med. Image Anal. **51**, 21–45 (2019)
9. Kim, R.J., et al.: Relationship of MRI delayed contrast enhancement to irreversible injury, infarct age, and contractile function. Circulation **100**(19), 1992–2002 (1999)
10. Moccia, S., et al.: Development and testing of a deep learning-based strategy for scar segmentation on CMR-LGE images. Magn. Reson. Mater. Phys. Biol. Med **32**(2), 187–195 (2019)
11. Oktay, O., et al.: Attention U-Net: learning where to look for the pancreas. arXiv preprint arXiv:1804.03999 (2018)
12. Organization, W.H.: Cardiovascular diseases (CVDs) (2017). https://www.who.int/news-room/fact-sheets/detail/cardiovascular-diseases-(cvds). Accessed 04 June 2020
13. Positano, V., et al.: A fast and effective method to assess myocardial necrosis by means of contrast magnetic resonance imaging. J. Cardiovasc. Magn. Reson. **7**(2), 487–494 (2005)
14. Ronneberger, O., Fischer, P., Brox, T.: U-Net: convolutional networks for biomedical image segmentation. In: Navab, N., Hornegger, J., Wells, W.M., Frangi, A.F. (eds.) MICCAI 2015. LNCS, vol. 9351, pp. 234–241. Springer, Cham (2015). https://doi.org/10.1007/978-3-319-24574-4_28
15. de la Rosa, E., Sidibé, D., Decourselle, T., Leclercq, T., Cochet, A., Lalande, A.: Myocardial infarction quantification from late gadolinium enhancement MRI using top-hat transforms and neural networks. arXiv preprint arXiv:1901.02911 (2019)
16. Szegedy, C., Vanhoucke, V., Ioffe, S., Shlens, J., Wojna, Z.: Rethinking the inception architecture for computer vision. In: Proceedings of the IEEE Conference on Computer Vision and Pattern Recognition, pp. 2818–2826 (2016)
17. Wong, K.C.L., Moradi, M., Tang, H., Syeda-Mahmood, T.: 3D segmentation with exponential logarithmic loss for highly unbalanced object sizes. In: Frangi, A.F., Schnabel, J.A., Davatzikos, C., Alberola-López, C., Fichtinger, G. (eds.) MICCAI 2018. LNCS, vol. 11072, pp. 612–619. Springer, Cham (2018). https://doi.org/10.1007/978-3-030-00931-1_70
18. Zabihollahy, F., White, J.A., Ukwatta, E.: Myocardial scar segmentation from magnetic resonance images using convolutional neural network. In: Medical Imaging 2018: Computer-Aided Diagnosis, vol. 10575, p. 105752Z. International Society for Optics and Photonics (2018)
19. Zhao, S., Wang, Y., Yang, Z., Cai, D.: Region mutual information loss for semantic segmentation. In: Advances in Neural Information Processing Systems, pp. 11117–11127 (2019)

20. Zhuang, X.: Multivariate mixture model for cardiac segmentation from multi-sequence MRI. In: Ourselin, S., Joskowicz, L., Sabuncu, M.R., Unal, G., Wells, W. (eds.) MICCAI 2016. LNCS, vol. 9901, pp. 581–588. Springer, Cham (2016). https://doi.org/10.1007/978-3-319-46723-8_67
21. Zhuang, X.: Multivariate mixture model for myocardial segmentation combining multi-source images. IEEE Trans. Pattern Anal. Mach. Intell. **41**(12), 2933–2946 (2019)

Dual Attention U-Net for Multi-sequence Cardiac MR Images Segmentation

Hong Yu[1], Sen Zha[1], Yubin Huangfu[1], Chen Chen[2], Meng Ding[3], and Jiangyun Li[1(✉)]

[1] School of Automation and Electrical Engineering,
University of Science and Technology Beijing, Beijing, China
{g20198754,g20198675}@xs.ustb.edu.cn, huangfuyb@foxmail.com,
leejy@ustb.edu.cn
[2] Department of Electrical and Computer Engineering,
University of North Carolina at Charlotte, Charlotte, USA
chen.chen@uncc.edu
[3] Scoop Medical, Houston, TX, USA
meng.ding@okstate.edu

Abstract. Myocardial pathology segmentation in cardiac magnetic resonance (CMR) is significant in the diagnosis for patients suffering from myocardial infarction (MI). Therefore, accurate and automatic segmentation method is highly desired in clinical practice. To better accomplish this segmentation task, we propose a modified U-net architecture named Dual Attention U-net. In this network, we use U-net as the baseline and embed a dual-branch attention module in it. One of the branches provides channel attention via emphasizing feature association among different channel maps, while the other branch provides spatial attention which adaptively aggregates the features at relative positions regardless of their distances in a weighted manner. Experiments show that both of these modules have effectively improved the segmentation performance. In addition, we have adopted data processing and augmentation methods to further improve the segmentation quality. Our model is evaluated on the public dataset from the MyoPS 2020 challenge, which consists of three sequences of cardiac MR images (bSSFP, LGE, and T2-weighted) from 45 patients. Our method achieves the Dice score of 63.5 (scar) and 68.8 (scar and edema) in the final test set.

Keywords: Cardiac magnetic resonance · Dual Attention U-net · Medical image segmentation

1 Introduction

A comprehensive and multidimensional cardiac view can be showed by cardiac magnetic resonance image (MRI). In the clinical treatment of myocardial infarction, Cardiac MRI is beneficial to the assessment of the cardiac function [4]. Each of the Cardiac MRI modalities contains different priorities. The late gadolinium enhancement (LGE) sequence is able to visualize the area of infarcted

© Springer Nature Switzerland AG 2020
X. Zhuang and L. Li (Eds.): MyoPS 2020, LNCS 12554, pp. 118–127, 2020.
https://doi.org/10.1007/978-3-030-65651-5_11

myocardium, the T2-weighted CMR emphasizes the myocardial edema regions, and the balanced-Steady State Free Precession (bSSFP) CMR provides specific information of cardiac boundaries. Thus, integrating multi-sequence CMR contributes to capture abundant and available pathological and morphological information of the myocardium [16]. Accurate delineation of scar and edema regions in CMR is significant in the analysis and diagnosis of patients suffering myocardial infarction.

In the clinical application, segmentation of particular organs or tissues in medical imaging still heavily depends on manual operation. However, artificial delineation is generally time-wasting, humdrum and subjective [16]. An automatic and robust segmentation method is in great request. Recently, deep learning has obtained remarkable success over various computer vision tasks [5,8]. Specifically, U-shape network structure [9] plays a significant role in medical image segmentation, and numerous variants have been developed. Chen et al. [2] proposed two-task recursive attention model to segment the left atrium and the atrial scars simultaneously. Combining histogram matching method, Liu et al. [7] achieved effective augmentation of the training data and acquired ventricles and myocardium segmentation result. Wang et al. [12] presented SK-Unet to segment left ventricle, right ventricle and left ventricular myocardium, which embed attention modules in different stages of the structure based on U-net. Furthermore, there are also several works aiming in the segmentation of cardiac anatomical structures in CMR [1,6,10,13,14].

In this paper, we present an novel algorithm based on modified U-net to segment myocardial edema, scar and other cardiac tissues automatically. The data is from the challenge of myocardial pathology segmentation combining multi-sequence CMR (MyoPS 2020), which consists of 45 cases of multi-sequence CMR. Data pre-processing and augmentation is employed to facilitate the training process and avoid overfitting of the model. The boundary of the pathological region in the heart showed in MRI is blurred and difficult to distinguish, which is quite a challenge for existing segmentation network. Previous research has established that contextual information and spatial information benefits pixel-wise classification and improves the recognition of object boundaries. Inspired by [3,11], the dual attention module [3] is plugged into the sub-module of modified U-net architecture. The spatial attention aggregates pixel-wise contextual information adaptively. In addition, the channel attention module can obtain the connection between different feature maps of channel dimension. Both modules embedded in U-net aid to restore the border of target and improve segmentation accuracy. With this approach, we get the final segmentation results of scar and edema.

2   Method

Our proposed model is based on the classical U-net [9], which is effective in medical image segmentation tasks. We have made considerable improvements based on U-net to adapt to the segmentation of cardiac MR images. Our fully automatic segmentation algorithm can precisely identify myocardial scars and edema regions in an end-to-end manner. In addition, we will also introduce the data processing and data augmentation strategies that are used in our experiments.

2.1 Data Processing

Data Preprocessing. The MRI dataset used to evaluate our proposed network is provided by the MyoPS 2020 challenge. It contains 45 cases of three-sequence CMR. Each CMR consists of three different modalities(e.g., LGE, T2 and bSSFP). Images of different modalities in the dataset have been aligned into a common space and re-sampled into the same spatial resolution by using the *MvMM method* [15]. To fully utilize of the information of the three modalities, we empirically treat the three sequences as three channels and stack them into one image. Since this competition aims to distinguish myocardial infarction area from healthy tissue, we remove the images without lesions from the dataset. Ultimately, we obtain 97 three-channel images as the training data. The pixel values of the MRIs are extremely large, we therefore normalize the input images to the distribution of zero mean and variance of one within the heart area. Besides, we also have tried to apply the contrast limited adaptive histogram equalization (CLAHE) and histogram matching to get relatively consistent intensity among the three modality images.

Data Augmentation. Due to the limited training images, data augmentation becomes extremely important to improve the performance of the model. In the experiments, we found non-rigid transformations (e.g., ElasticTransform, GridDistortion and OpticalDistortion) are effective techniques to gain a better performance in medical image segmentation. The visualization of three non-rigid transformations is shown in Fig. 1. Consequently, our data augmentation methods include flipping, cropping, rotation, brightness and contrast shift and non-rigid transformations.

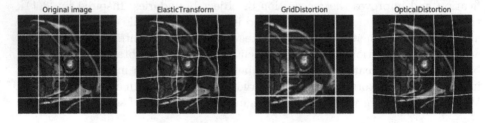

Fig. 1. A visualization of the original image and counterparts generated by three kinds of non-rigid transformations. We add white grids to the images for a better comparison.

2.2   Dual Attention U-Net Architecture

The proposed Dual Attention U-net is a segmentation network composed of encoder and decoder. The shortcut connection combines the same size feature maps in the encoder and decoder. At the end of the encoder, we embed the dual attention module [3] to capture the global context information on feature maps and enhance specific feature maps among different channels.

The network architecture is illustrated in Fig. 2. The inputs of the network are three sequences of cardiac MR images while the output is the probability that each pixel is classified as background, edema, scar, etc. The left side of the network is the encoder part, which is a classical convolutional architecture. This architecture extracts feature information by repeatedly applying two 3 × 3 convolutions and each followed by a batch normalization layer, a rectified linear unit (ReLU) and a 2 × 2 max pooling operation. The pooling layer uses a stride of 2 for downsampling, and the number of feature channels are doubled at each feature map downsampling. The left and right sides of the network are symmetrical. In the decoder we replace the max pooling with deconvolution in order to upsample the feature maps. The feature maps of the same size in the encoder and decoder are connected by shortcuts and concatenated in the channel dimension, leading to the fusion of features from shallow and deep convolution layers.

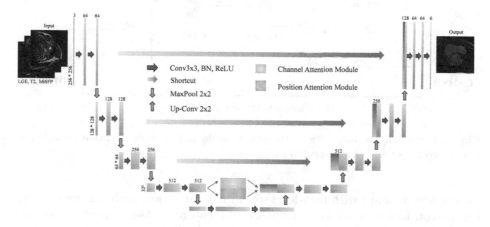

Fig. 2. The overall structure of our proposed Dual Attention U-net.

As shown in the network architecture, we insert the **channel attention** module (CAM) to selectively emphasize feature association among different channel maps and the **position attention** module (PAM) to capture the long-range dependency on feature maps. In Fig. 3, these two modules are illustrated in detail. In the PAM, given an existing feature map $I \subset R^{C \times H \times W}$, we first use two 1 × 1 convolutions to generate corresponding new feature maps Q and K respectively, where $Q, K \in R^{C' \times H \times W}$. Then we reshape them to $R^{C' \times N}$, where

$N = H \times W$ means the number of pixels in each channel. After that, the result of multiplying the transposed Q and K matrices is sent to the softmax layer to calculate the spatial attention map $S \in R^{N \times N}$. At the same time, we feed the feature map I into another 1×1 convolution to generate a new feature map $V \in R^{C \times H \times W}$ and reshape it to $R^{C \times N}$. Similarly, we perform a matrix multiplication between V and the transpose of S and reshape the result O to $R^{C \times H \times W}$. In the CAM, except for deprecating the 1×1 convolution, the other steps are basically the same.

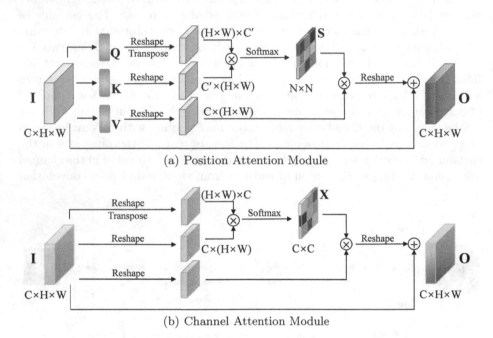

(a) Position Attention Module

(b) Channel Attention Module

Fig. 3. The details of the position attention module and channel attention module are illustrated in (a) and (b), respectively.

For feature maps with high-level semantic information, each feature channel can be considered as a specific response to a category/class. Therefore, we use the CAM to take advantage of the interdependence between the feature map channels, which is very helpful for pixel-level classification tasks. On the other hand, capturing long-range dependence is of great importance. In convolutional neural networks, the long-range dependence is captured by the large receptive field formed by stacking convolutional layers. Therefore, by repeatedly stacking convolutional layers and spreading the data layer by layer is possible to capture long-range dependencies. However, this method is computationally inefficient and leads to difficulties in parameter optimization. To this end, we use the PAM to accomplish this goal.

2.3 Post Processing

Image post processing is adopted to refine the results of cardiac pathology segmentation. Firstly, the prior knowledge that the pathological area appears around myocardium is applied to obtain more reasonable segmentation. Therefore, we remove the obviously wrong segmentation results which are outside the target area. From the training data, we observed that the labeled pathology region of each sample is basically a whole connected area, that inspired us to remove the unconnected segmentation results of pathology regions which are usually not reasonable and fill with adjacent category pixels under a predefined threshold. This threshold is chosen for each experiment independently by optimizing the mean Dice on the MyoPS 2020 training cases. The specific effect of post-processing is shown in Fig. 4.

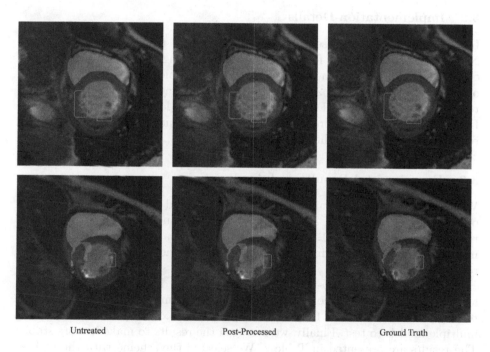

Untreated   Post-Processed   Ground Truth

Fig. 4. Visualization of post-processing effects.

3 Experiments and Results

3.1 Data and Evaluation Metrics

The dataset provided by MyoPS 2020 contains 45 cases of multi-sequence CMR, and each case refers to a patient with three sequence CMR, i.e., LGE, T2 and bSSFP CMR. The first 25 cases are used as training data while the rest are reserved as test data. In the training set, we separated the last 5 cases for validation. The labels for myocardial pathology segmentation include left ventricular (LV) blood pool (labelled 500), right ventricular blood pool (600), LV normal myocardium (200), LV myocardial edema (1220) and LV myocardial scars (2221). However, the evaluation of the test data will be focused on the myocardial pathology segmentation, i.e., scars and edema.

The evaluation metrics used in our experiment include the Dice coefficient (Dice) and Hausdorff distance (HD). Dice score is widely used for evaluating medical image segmentation quality and it measures the degree of overlap between the predicted segmentation map and ground truth. HD measures the maximum degree of mismatch between the ground truth and predicted object boundaries.

3.2 Implementation Details

In our experiments, we used a batch size of 8 and trained the model on an Nvidia TITAN RTX GPU for 500 epochs. Considering the segmentation object only occupies a small area in the center of the whole image, we cropped the original image into a 256×256 voxel to remove the black areas (i.e., background regions). We adopt the SGD optimizer with an initial learning rate of 0.001. The Dice loss is employed to train the network and L2 norm is applied for model regularization with a weight decay rate of 10^{-5}.

3.3 Experimental Results and Analysis

When evaluating on the validation set, we conducted experiments respectively on the use of PAM, CAM, and both of them simultaneously. The experimental results are listed in Table 1. Comparing to the classical U-Net, no matter which module we use, it can bring performance improvements to the segmentation results.

In the experiment, non-rigid transformation and data post-processing have a positive effect. In addition, we utilized a test augmentation technique when testing the model, that is, flipping the input image axes respectively and using multiple models to test. Finally, we averaged the results to make it more stable. The results are presented in Table 2. We selected the scheme with the highest score on the validation set for the prediction of 20 cases in the test set, and sent the results to the competition organizer for evaluation. Our final performance on the test set is shown in Table 3. Similarly, the effectiveness of PAM, CAM, and both of them was verified respectively in the test set by test tool provided by the sponsor [15]. After applying the PAM and CAM on the experiment, the final dice score is further improved.

The computational complexity comparison (Table 4) between U-net and our Dual Attention U-net shows that the additional channel and position attention modules only bring a slight increase in the computational cost, which has a very minor effect on the speed of reasoning.

Table 1. Performance comparison on the validation set.

(a) Comparison of Dice Score

Method	Dice Score(%) ↑				
	Myo	LV	RV	Scar	Scar+Edema
U-net	73.24	88.22	76.66	60.88	68.16
U-net + CAM	73.82	88.83	77.47	64.45	71.49
U-net + PAM	**75.61**	**89.45**	**78.02**	64.04	71.69
U-net + CAM +PAM	75.18	89.19	77.24	**64.82**	**72.39**

(b) Comparison of Hausdorff Distance

Method	Hausdorff Distance(mm) ↓				
	Myo	LV	RV	Scar	Scar+Edema
U-net	23.40	7.53	12.74	24.08	31.09
U-net + CAM	18.13	7.15	7.89	18.03	21.12
U-net + PAM	17.78	7.00	**5.12**	17.61	24.98
U-net + CAM +PAM	**16.81**	**6.88**	5.49	**16.35**	**20.92**

Table 2. Performance of multiple strategies on the validation set. We use U-net + CAM + PAM as the base model to obtain various performances using non-rigid transformation (NR), data post-processing (PP) and test augmentation (TA).

Method	Dice score (%) ↑		Hausdorff distance (mm) ↓	
	Scar	Scar+Edema	Scar	Scar+Edema
Base	55.9	68.71	28.82	35.96
Base + NR	64.33	71.58	24.64	30.94
Base + NR + PP	64.74	72.19	18.60	22.62
Base + NR + PP + TA	**64.82**	**72.39**	**16.35**	**20.92**

This competition focuses on the segmentation of myocardial pathological regions. We selected three cases for visualization as shown in Fig. 5. The visualization can further illustrate that the segmentation results has been significantly improved. For small areas such as scars and edema, PAM can achieve more accurate segmentation by capturing global contextual correlations. Also, as shown in the visualization, over-segmentation occurred at a distance from the lesion, while U-net with PAM effectively alleviates this phenomenon. Meanwhile, CAM can also make the channel respond to specific categories and improve the performance of pathological region segmentation.

Table 3. Performance of our approach on the final test set.

Method	Dice score_mean ↑		Dice score_std ↓	
	Scar	Scar+Edema	Scar	Scar+Edema
U-net	0.6134	0.6755	**0.2701**	0.1475
U-net + CAM	0.6282	0.6825	0.2861	0.1495
U-net + PAM	0.6294	0.6806	0.2803	**0.1464**
U-net + CAM + PAM	**0.6345**	**0.6880**	0.2899	0.1484

Table 4. Computational complexity analysis.

Model	FlOPs (G)	Parameter (M)
U-net	46.177	31.04
U-net + CAM + PAM	46.513	31.37

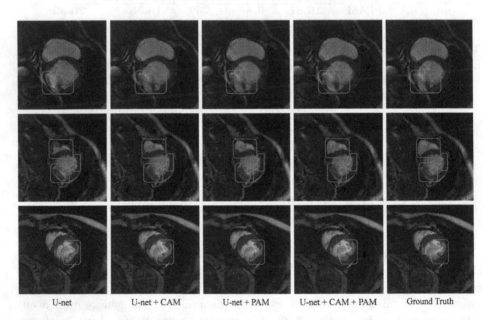

U-net   U-net + CAM   U-net + PAM   U-net + CAM + PAM   Ground Truth

Fig. 5. The segmentation results. From left to right are the original U-net, U-net with CAM, U-net with PAM, U-net with both CAM and PAM, and the ground truth.

4   Conclusion

In this paper, we proposed a Dual Attention U-net for myocardial pathology segmentation in cardiac magnetic resonance images, which adaptively integrates local semantic features using the self-attention mechanism. Specifically, we used two branches of attention modules to capture global dependencies in the spatial and channel dimensions respectively. The experiments show that the Dual

Attention U-net yields more precise segmentation results for myocardial pathology segmentation than the original U-net. Our approach and results analysis can also provide useful network design insights to aided diagnoses for clinical cardiac surgeon.

References

1. Campello, V.M., Martinisla, C., Izquierdo, C., Petersen, S.E., Ballester, M.A.G., Lekadir, K.: Combining multi-sequence and synthetic images for improved segmentation of late gadolinium enhancement cardiac MRI. arXiv-Image and Video Processing (2019)
2. Chen, J., et al.: Multiview two-task recursive attention model for left atrium and atrial scars segmentation, pp. 455–463 (2018)
3. Fu, J., et al.: Dual attention network for scene segmentation, pp. 3146–3154 (2019)
4. Han, W.K., Farzaneh-Far, A., Kim, R.J.: Cardiovascular magnetic resonance in patients with myocardial infarction: current and emerging applications. J. Am. Coll. Cardiol. **55**(1), 1–16 (2009)
5. He, K., Zhang, X., Ren, S., Sun, J.: Deep residual learning for image recognition, pp. 770–778 (2016)
6. Khened, M., Alex, V., Krishnamurthi, G.: Fully convolutional multi-scale residual DenseNets for cardiac segmentation and automated cardiac diagnosis using ensemble of classifiers. Med. Image Anal. **51**, 21–45 (2018)
7. Liu, Y., Wang, W., Wang, K., Ye, C., Luo, G.: An automatic cardiac segmentation framework based on multi-sequence MR image. arXiv-Image and Video Processing (2019)
8. Long, J., Shelhamer, E., Darrell, T.: Fully convolutional networks for semantic segmentation, pp. 3431–3440 (2015)
9. Ronneberger, O., Fischer, P., Brox, T.: U-Net: convolutional networks for biomedical image segmentation. CoRR abs/1505.04597 (2015). http://arxiv.org/abs/1505.04597
10. Shi, Z., et al.: Bayesian VoxDRN: a probabilistic deep Voxelwise dilated residual network for whole heart segmentation from 3D MR images, pp. 569–577 (2018)
11. Wang, X., Girshick, R., Gupta, A., He, K.: Non-local neural networks, pp. 7794–7803 (2018)
12. Wang, X., et al.: SK-Unet: an improved u-net model with selective kernel for the segmentation of multi-sequence cardiac MR. arXiv-Image and Video Processing (2020)
13. Xiong, Z., Fedorov, V.V., Fu, X., Cheng, E., Macleod, R.S., Zhao, J.: Fully automatic left atrium segmentation from late gadolinium enhanced magnetic resonance imaging using a dual fully convolutional neural network. IEEE Trans. Med. Imaging **38**(2), 515–524 (2019)
14. Zabihollahy, F., White, J.A., Ukwatta, E.: Myocardial scar segmentation from magnetic resonance images using convolutional neural network. In: Computer-Aided Diagnosis (2018)
15. Zhuang, X.: Multivariate mixture model for myocardial segmentation combining multi-source images. IEEE Trans. Pattern Anal. Mach. Intell. **41**(12), 2933–2946 (2019)
16. Zhuang, X.: Multivariate mixture model for cardiac segmentation from multi-sequence MRI. In: International Conference on Medical Image Computing & Computer-assisted Intervention (2016)

Accurate Myocardial Pathology Segmentation with Residual U-Net

Altunok Elif$^{(\boxtimes)}$ and Oksuz Ilkay

Computer Engineering Department, Istanbul Technical University, Istanbul, Turkey
altunokelif@gmail.com

Abstract. Accurate assessment of myocardial viability in multi-sequence cardiac magnetic resonance (CMR) images is desired to automate disease diagnosis. To classify myocardial pathology automatic segmentation methods are necessary. In this paper we propose to use an automatic segmentation for each slice in the short-axis view with convolutional neural network architecture based on U-Net. We compare performances of two different networks to segment myocardial pathologies. The best performance is obtained by using the U-net convolutional neural network architecture built from residual units trained by augmentation operations, showing that it is a practical approach for segmentation. The network performances are assessed on MyoPS 2020 challenge dataset consists of three-sequence CMR images from 45 patients. A five-fold cross-validation strategy is utilized to assess performance of the proposed method.

Keywords: Cardiac MR · Scar · Edema · LGE · Image segmentation · Convolutional neural networks

1 Introduction

Cardiovascular diseases (CVD), which cause of death approximately 17.9 million people each year, are the number one cause of death globally [1]. An expert analyzes cardiac MR images for diagnosing the disease, though the analysis process takes time, and the expert has limited time for this task. This is time-consuming, tiring, and prone to subjective errors. Furthermore, the speed of diagnosis is very important in critical situations in these diseases. Accurate and early detection is of great importance for individuals with this disease or individuals in the risk group. Since cardiovascular diseases are a problem throughout society, developments in this area directly affect people's health. For all these reasons, it is very important to manage this process automatically. The segmentation of cardiac MRI data plays an important role in this process.

The cardiac shape can change drastically due to variations in anatomy from patient to patient. Variations related to acquisition such as different angles and poor quality shots of CMR images can be found in the data. Therefore, it is essential to develop algorithms that can address these variations in data. In

© Springer Nature Switzerland AG 2020
X. Zhuang and L. Li (Eds.): MyoPS 2020, LNCS 12554, pp. 128–137, 2020.
https://doi.org/10.1007/978-3-030-65651-5_12

this paper, we describe our method for myocardial pathology segmentation of multi-sequence CMR images using neural network architectures based on U-Net. Accurate segmentation of myocardial edema and scars is crucial in making the diagnosis automatically. We compare the performance of models trained with two different architectures U-Net and Residual U-Net.

This study is our entry for MyoPS 2020 challenge, which aims the myocardial pathology segmentation from multi-sequence CMR. Although the challenge only evaluates the results of the pathology segmentation, i.e., scars, and edema of left ventricle myocardium, we will also discuss in this paper our results with ventricular blood pools.

2   Background

Segmentation of myocardial region has been a key challenge and machine learning techniques have heavily influenced the trends. Earlier approaches are developed based on semi-automated and atlas-based to perform the segmentation of cardiac MR images (CMR) [11,12]. Recently, deep learning plays a great role in the semantic segmentation of CMR images. Convolutional networks have applied on segmentation task for a long time, but their success was limited due to both the size of the available training sets and the size of the networks considered [8]. However, the advancement of deep learning has also led to success in segmentation. More recently, many approaches based on deep learning overcome the performance of classical approaches for cardiac image segmentation. Tran used a fully convolutional neural network for cardiac segmentation in short-axis MRI [15]. Recurrent Neural Network (RNN) approach is applied for cardiac segmentation in MR images successively [12]. The U-net architecture of Ronneberger et al. has been widely used in the area of medical image segmentation [13]. ResNet architecture of has achieved remarkable success in computer vision tasks [4]. This motivated research on using ResNet for segmentation architectures to assess cardiac health and identify certain pathologies [7]. Karim et al. [6] developed methodologies that detect and quantify infarct in late gadolinium enhancement magnetic resonance images (LGE-MRI) for actual clinical practice. Zabihollahy et al. [19] used convolutional neural network (CNN) to segmentation of the myocardial fibrosis or scar in LGE-MRI. The architectures has not been the only focus for performance improvement in cardiac MRI segmentation. Several loss functions, such as weighted cross-entropy [5], Dice loss [10], weighted Dice loss [18], focal loss Sander et al. [14], and deep supervision loss [3] have been investigated to increase the alignment with ground truth annotations. In recent years, multiple challenges on the segmentation task have also been opened for advancing the state of the art for segmentation tasks [2], but they failed short to focus on regional pathology such as scar and edema. In this work, we showcase the performance of residual U-net architecture on myocardial pathology segmentation.

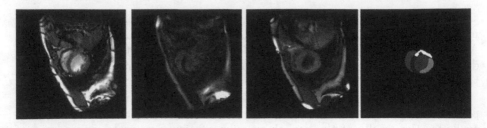

Fig. 1. An example of multi-sequence CMR image from training dataset. First image bSSFP, second image LGE, third image T2 CMR, and fourth image is associated ground truth segmentation. Myocardium region segmented into normal, edema, and scar regions, where dark gray represents the normal myocardium, gray represents the edema, and white represents the scar.

3   Methods

3.1   Dataset

In this study, the network performance assessed on myocardial pathology segmentation combining multi-sequence CMR Challenge for MyoPS 2020 dataset [20,21]. Training dataset consists of 25 cases having a different number of slices of multi-sequence CMR, i.e., late gadolinium enhancement (LGE), T2-weighted CMR which images the acute injury and ischemic regions, balanced-Steady State Free Precession (bSSFP) CMR, and all ground truth values for every single slice. The ground truth labels include left ventricular (LV) blood pool, right ventricular (RV) blood pool, LV normal myocardium, LV myocardial edema, LV myocardial scars and evaluation of the test data will only focus on myocardial pathology segmentation, i.e., scars and edema. The test dataset consists of 20 cases.

Accurate segmentation scar geometry in CMR images is essential for representing of patient-specific structural remodeling [16,17]. 2-dimensional LGE MRI is used to identify myocardial scar. Although a multi-sequence CMR data set provides us with rich and reliable information about the pathological and morphological information of the myocardium, the segmentation of scar and edema regions is hard to obtain and time-consuming in such a dataset. Moreover, the edema and scar may be concentrated in a small region and exhibit low intensity distinction, which makes them more difficult to extract from the rest of the anatomy. MyoPS 2020 dataset presents us a great challenge for segmenting edema and scar regions since the dataset directly collected from the clinic without any selection. In Fig. 1, we show an example input multi-sequence CMR image and its ground truth annotation from the training dataset.

3.2   U-Net Architecture

The U-Net architecture, a convolutional network for biomedical image segmentation is used to segmentation of the cardiac MR images. The network consists of a contracting (downsampling) and a expanding (upsampling) paths. In the contracting path, there are 3×3 convolutions, each followed by a rectified linear unit (ReLU) activation function and 2×2 max-pooling operation with stride 2. There is also dropout layer after first convolution at each downsampling step to prevent the architecture from overfitting. In the contracting step the number of feature channels doubles. In the expansive, there are upsampling of the feature map followed by a 2×2 convolution in all steps. At each step, the number of feature channels halves. There are also a concatenation with the feature map from the contracting path, and two 3×3 convolutions, each followed by a ReLU activation function. A 1×1 convolution also is used to map each feature vector to the number of classes at the final layer.

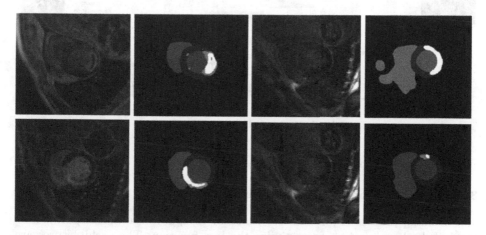

Fig. 2. From top to bottom, the example of CMR image segmentation results of the U-Net method, and our proposed method in the MyoPS 2020 test dataset. The left result is best in terms of segmentation of edema and scar and the right result is the worst-case scenario. Normal myocardium shown in dark gray, myocardial scars shown in white, and myocardial edema shown in gray in the predicted segmentation mask.

3.3   Residual U-Net Architecture

Residual U-Net architecture consists of three parts: encoding, bridge and decoding. We use a 5×5 zero padding, a 7×7 convolution with a stride of 2, a batch normalization layer, followed by a ReLU activation function, and a max-pooling operation before making operations in the encoding path. Each down box of the encoding portion is implemented with a downsampling block and two residual units. We use convolutions with a stride of 1 for residual units. In downsampling block, we use convolutions with a stride of 2. In the decoding path, each up box is implemented with an upsampling block and two residual units. In upsampling block, there are an upsampling layer and a convolution layer with a stride of 1. At each decoding unit, there is a concatenation with the feature maps from the corresponding encoding path. After last decoding unit, there is also one more upsampling block, two residual units, an upsampling, and a convolution operation. At final layer, there is a 1×1 convolutional layer having sigmoid activation function to generate a desired output for segmentation map. A detailed illustration of the architecture is provided in Fig. 3.

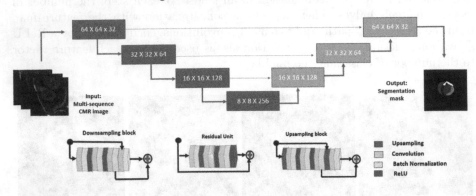

Fig. 3. Residual U-Net architecture

3.4   Loss Function and Evaluation Metric

The training process requires a loss function to update the model parameters through backpropagation to minimize the loss function. The architectures are trained with dice loss which is defined as follows:

$$L(y, \hat{y}) = -\frac{1}{M} \sum_{j=0}^{M} \sum_{i=0}^{N} (y_{ij} log(\hat{y}_{ij})) \tag{1}$$

where $\hat{y}$ is the predicted expected value, y is the observed value for each class N. Dice coefficient used for segmentation accuracy assessment by evaluating the overlap between the ground truth and the predicted area and defined as:

$$DC = \frac{2 A \cap B|}{|A| + |B|} \tag{2}$$

where A refers to manual segmentation and B refers to automated segmentation area. The $1 - DC$ is called soft-Dice loss function [9].

3.5   Implementation Details

We trained the Residual U-Net architecture using the Keras library on NVIDIA Quadro RTX 6000 GPU. The training takes approximately 2h for 750 epochs.

We uses the three sequences of CMR, i.e., bSSFP, LGE, and T2 CMR as input channels. The images form different subjects are in different image resolution. Therefore, we cropped the images with 256×256 pixels using the center point of the each slice. Then, we normalize the pixel values of each image both in the training set and test set to its pixel values having zero mean. We use ADAM optimizer with a learning rate set to 10^{-3}. Selecting the learning rate too small may result in a long training period, while a great value can lead to learning a very fast or unstable training process. To avoid this problem, the learning rate was reduced by multiplying 0.5 if validation dice loss does not improve for 10 epochs by using learning rate schedule. We also separately trained the U-Net and Residual U-Net networks on the five dataset folds. Each fold trained for 500 epochs. The learning rate set to 10^{-4} for U-Net and 10^{-3} for the Residual U-Net by using the same learning rate scheduler.

Data Augmentation. In case of insufficient data, it is necessary to increase the number of images by providing diversity in the dataset in order to prevent over-fitting. Data augmentation techniques include image dropping out of $1 - 5\%$ of the pixels, -90 to 90-degree rotation, horizontally flipping 50% of the images, and elastic transformations. The data augmentation method applied to both original images and ground truth masks. To provide expanded dataset, the training dataset was increased ten times (See Fig. 4).

Fig. 4. First image is for original MR image, second image is for original image with ground truth mask, third image is for augmented image, fourth image is for augmented ground truth mask on augmented image and last image is for augmented ground truth mask on its own in.

4 Experimental Results

4.1 K-Fold Cross-validation Results

We trained the U-Net and residual U-Net (RU-Net) architectures using 5-fold cross-validation in the training set. At each K-fold of the cross validation we use 20 images for training and 5 images for testing. We ensure each patient data to be in the test set using this scheme. Table 1 demonstrates the results for applying the trained networks to the validation set. Based on these results, we can observe that both networks with data augmentation produced the much better dice scores compared to training without augmentation. The same table illustrates the comparison between U-Net and Residual U-Net methods. Residual U-Net has significantly better results in segmenting the all regions in the multi-sequence CMRs. An example of CMR image segmentation validation result in 5th fold of the Residual U-Net method shown in Fig. 5.

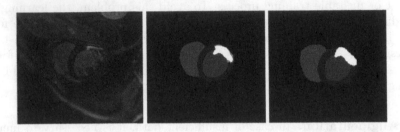

Fig. 5. Example of CMR image segmentation validation result in 5th fold of our proposed method in the MyoPS 2020 test dataset. First and second images are input image and corresponding ground truth segmentation from training dataset respectively. Third image is the predicted segmentation result. Normal myocardium shown in dark gray, myocardial scars shown in white, and myocardial edema shown in gray in the predicted segmentation mask.

4.2   Test Results

We trained both U-Net models, i.e. U-Net, and Residual U-Net on the entire training set using the best parameter setup from cross-validation. The U-Net model achieved the mean dice score of $0.888(\pm0.208)$ (myocardial scars), and $0.866(\pm0.225)$ (myocardial scars & edema) on the training set. We evaluated the U-Net model on the test set, and the model achieved the mean dice score of $0.524(\pm0.277)$ (myocardial scars), and $0.573(\pm0.180)$ (myocardial scars & edema) in the multi-sequence CMRs. The Residual U-Net achieved the mean dice score of $0.926(\pm0.115)$ (myocardial scars), and $0.911(\pm0.131)$ (myocardial scars & edema) on the training set. We evaluated the Residual U-Net model on the test set, and our model achieved the mean dice score of $0.565(\pm0.260)$ (myocardial scars), and $0.612(\pm0.160)$ (myocardial scars & edema) in the multi-sequence CMRs. Experiments show that our method achieved better performance than the basic U-Net method for the segmenting of myocardial pathology. Figure 2 shows the best and the worst visual segmentation results of applying both networks in terms of edema and scar regions of the same slice from one specific case in the test dataset. The predicted mask also includes RV, LV blood pools, and normal myocardium region.

Table 1. Fold validation results for the left ventricular blood pool (LV), right ventricular blood pool (RV), LV normal myocardium (Myo), LV myocardial edema (edema), LV myocardial scars (scars), and multi-class in terms of dice score and standard deviation in the CMR segmentation task. No-aug shows the results without data augmentation. All results are multiplied by 1000 and the bold font highlights the best mean results.

U-Net

Fold	LV	RV	Myo	Edema	Scars	Multi-class
1	822 ± 258	082 ± 005	629 ± 239	017 ± 000	305 ± 158	471 ± 108
2	840 ± 353	790 ± 409	680 ± 339	427 ± 264	223 ± 149	658 ± 248
3	863 ± 244	827 ± 267	616 ± 193	263 ± 107	411 ± 171	661 ± 160
4	807 ± 334	697 ± 366	637 ± 317	151 ± 073	447 ± 301	621 ± 224
5	382 ± 794	851 ± 258	639 ± 258	296 ± 150	535 ± 252	614 ± 239
Mean	743 ± 397	649 ± 261	640 ± 269	231 ± 119	384 ± 206	605 ± 196
No-aug	534 ± 547	366 ± 435	180 ± 179	109 ± 126	188 ± 249	388 ± 255

RU-Net

Fold	LV	RV	Myo	Edema	Scars	Multi-class
1	867 ± 204	843 ± 285	678 ± 211	119 ± 108	368 ± 174	637 ± 147
2	893 ± 199	880 ± 222	770 ± 208	354 ± 148	310 ± 118	699 ± 163
3	825 ± 209	751 ± 237	623 ± 178	283 ± 100	366 ± 138	630 ± 135
4	870 ± 256	837 ± 249	763 ± 204	261 ± 098	585 ± 206	716 ± 165
5	885 ± 238	863 ± 272	699 ± 242	336 ± 138	608 ± 231	728 ± 202
Mean	**868 ± 221**	**835 ± 253**	**706 ± 208**	**271 ± 118**	**448 ± 173**	**682 ± 163**
No-aug	701 ± 533	637 ± 548	465 ± 461	134 ± 158	181 ± 200	510 ± 323

5 Discussion and Conclusion

In this paper, a residual U-Net based architectures for the segmenting of myocardial pathology have been used and assessed on MyoPS 2020 dataset. We compare the performance of the U-Net and Residual U-Net architectures. The model based on deep residual network outperformed the performance of U-Net. Some test cases have been challenging to segment for the proposed technique due to incosistent intensity patterns in the edges. One avenue of improvement could be to incorporate temporal information from multiple temporal slices. Further investigation will be devoted to alleviate the problem of regional imbalance by advancing the loss functions.

Acknowledgement. This research was funded by 2232 International Fellowship for Outstanding Researchers Program of the Scientific and Technological Research Council of Turkey (TÜBİTAK) grant number 118C353.

References

1. World health organization (2017). https://www.who.int/news-room/fact-sheets/detail/cardiovascular-diseases-(cvds)
2. Bernard, O., et al.: Deep learning techniques for automatic MRI cardiac multi-structures segmentation and diagnosis: is the problem solved? IEEE Trans. Med. Imaging **37**(11), 2514–2525 (2018)
3. Chen, S., Ma, K., Zheng, Y.: Med3D: transfer learning for 3D medical image analysis. arXiv preprint arXiv:1904.00625 (2019)
4. He, K., Zhang, X., Ren, S., Sun, J.: Deep residual learning for image recognition. In: Proceedings of the IEEE Conference on Computer Vision and Pattern Recognition, pp. 770–778 (2016)
5. Jang, Y., Hong, Y., Ha, S., Kim, S., Chang, H.-J.: Automatic segmentation of LV and RV in cardiac MRI. In: Pop, M., et al. (eds.) STACOM 2017. LNCS, vol. 10663, pp. 161–169. Springer, Cham (2018). https://doi.org/10.1007/978-3-319-75541-0_17
6. Karim, R., et al.: Evaluation of state-of-the-art segmentation algorithms for left ventricle infarct from late gadolinium enhancement MR images. Med. Image Anal. **30**, 95–107 (2016)
7. Kerfoot, E., Clough, J., Oksuz, I., Lee, J., King, A.P., Schnabel, J.A.: Left-ventricle quantification using residual U-Net. In: Pop, M., et al. (eds.) STACOM 2018. LNCS, vol. 11395, pp. 371–380. Springer, Cham (2019). https://doi.org/10.1007/978-3-030-12029-0_40
8. LeCun, Y., et al.: Backpropagation applied to handwritten zip code recognition. Neural Comput. **1**(4), 541–551 (1989)
9. Milletari, F., Navab, N., Ahmadi, S., V-Net: fully convolutional neural networks for volumetric medical image segmentation. In: Proceedings of the 2016 Fourth International Conference on 3D Vision (3DV), pp. 565–571 (2016)
10. Milletari, F., Navab, N., Ahmadi, S.A.: V-Net: fully convolutional neural networks for volumetric medical image segmentation. In: 2016 Fourth International Conference on 3D Vision (3DV), pp. 565–571. IEEE (2016)
11. Petitjean, C., Dacher, J.N.: A review of segmentation methods in short axis cardiac MR images. Med. Image Anal. **15**(2), 169–184 (2011)
12. Poudel, R.P.K., Lamata, P., Montana, G.: Recurrent fully convolutional neural networks for multi-slice MRI cardiac segmentation. In: Zuluaga, M.A., Bhatia, K., Kainz, B., Moghari, M.H., Pace, D.F. (eds.) RAMBO/HVSMR -2016. LNCS, vol. 10129, pp. 83–94. Springer, Cham (2017). https://doi.org/10.1007/978-3-319-52280-7_8
13. Ronneberger, O., Fischer, P., Brox, T.: U-Net: convolutional networks for biomedical image segmentation. In: Navab, N., Hornegger, J., Wells, W.M., Frangi, A.F. (eds.) MICCAI 2015. LNCS, vol. 9351, pp. 234–241. Springer, Cham (2015). https://doi.org/10.1007/978-3-319-24574-4_28
14. Sander, J., de Vos, B.D., Wolterink, J.M., Išgum, I.: Towards increased trustworthiness of deep learning segmentation methods on cardiac MRI. In: Medical Imaging 2019: Image Processing, vol. 10949, p. 1094919. International Society for Optics and Photonics (2019)
15. Tran, P.V.: A fully convolutional neural network for cardiac segmentation in short-axis MRI. arXiv preprint arXiv:1604.00494 (2016)
16. Ukwatta, E., et al.: Image-based reconstruction of three-dimensional myocardial infarct geometry for patient-specific modeling of cardiac electrophysiology. Med. Phys. **42**(8), 4579–4590 (2015)

17. Vadakkumpadan, F., Gurev, V., Constantino, J., Arevalo, H., Trayanova, N.: Modeling of whole-heart electrophysiology and mechanics: toward patient-specific simulations. In: Kerckhoffs, R. (ed.) Patient-Specific Modeling of the Cardiovascular System, pp. 145–165. Springer, New York (2010)

18. Yang, X., Bian, C., Yu, L., Ni, D., Heng, P.-A.: Class-balanced deep neural network for automatic ventricular structure segmentation. In: Pop, M., et al. (eds.) STACOM 2017. LNCS, vol. 10663, pp. 152–160. Springer, Cham (2018). https://doi.org/10.1007/978-3-319-75541-0_16

19. Zabihollahy, F., White, J.A., Ukwatta, E.: Myocardial scar segmentation from magnetic resonance images using convolutional neural network. In: Medical Imaging 2018: Computer-Aided Diagnosis, vol. 10575, p. 105752Z. International Society for Optics and Photonics (2018)

20. Zhuang, X.: Multivariate mixture model for cardiac segmentation from multi-sequence MRI. In: Ourselin, S., Joskowicz, L., Sabuncu, M.R., Unal, G., Wells, W. (eds.) MICCAI 2016. LNCS, vol. 9901, pp. 581–588. Springer, Cham (2016). https://doi.org/10.1007/978-3-319-46723-8_67

21. Zhuang, X.: Multivariate mixture model for myocardial segmentation combining multi-source images. IEEE Trans. Pattern Anal. Mach. Intell. 41(12), 2933–2946 (2019)

Stacked and Parallel U-Nets with Multi-output for Myocardial Pathology Segmentation

Zhou Zhao$^{(\boxtimes)}$, Nicolas Boutry, and Élodie Puybareau

EPITA Research and Development Laboratory (LRDE), Le Kremlin-Bicêtre, France
{zz,nicolas.boutry,elodie.puybareau}@lrde.epita.fr

Abstract. In the field of medical imaging, many different image modalities contain different information, helping practitionners to make diagnostic, follow-up, etc. To better analyze images, mixing multi-modalities information has become a trend. This paper provides one cascaded UNet framework and uses three different modalities (the late gadolinium enhancement (LGE) CMR sequence, the balanced- Steady State Free Precession (bSSFP) cine sequence and the T2-weighted CMR) to complete the segmentation of the myocardium, scar and edema in the context of the MICCAI 2020 myocardial pathology segmentation combining multi-sequence CMR Challenge dataset (MyoPS 2020). We evaluate the proposed method with 5-fold-cross-validation on the MyoPS 2020 dataset.

Keywords: Deep learning · Myocardial pathology · Segmentation · UNet

1 Introduction

The assessment of myocardial viability is essential for diagnosis and follow-up of patients suffering from myocardial infarction (MI) [16,17]. However, many different images modalities in the field of medical imaging are available and are complementary. Late gadolinium enhancement (LGE) cardiac magnetic resonance (CMR) sequence which visualizes MI, T2-weighted CMR (imaging the acute injury and ischemic regions) and balanced-Steady State Free Precession (bSSFP) cine sequence (which captures cardiac motions and presents clear boundaries) are examples of such imaging modalities. Therefore, making a better use of the information in these different modalities has become a research focus. In recent years, many semi-automated and automated methods have been proposed for multi-modal medical image segmentation using deep learning-based methods, such as convolutional neural networks (CNNs) [8] and fully convolutional networks (FCNs) [9] especially the U-Net architecture [11]. For example, Guo [3,4] proposed a conceptual image fusion architecture for supervised biomedical image analysis. They designed and implemented an image segmentation system based on deep CNNs to contour the lesions of soft tissue sarcomas using multimodal images by fusing the information derived from different modalities.

© Springer Nature Switzerland AG 2020
X. Zhuang and L. Li (Eds.): MyoPS 2020, LNCS 12554, pp. 138–145, 2020.
https://doi.org/10.1007/978-3-030-65651-5_13

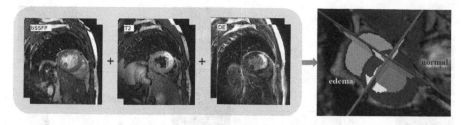

Fig. 1. Myocardial pathology, the picture is from MyoPS2020 challenge (http://www.sdspeople.fudan.edu.cn/zhuangxiahai/0/MyoPS20/data1.html).

Although we can use multi-modal information to improve the myocardial pathology segmentation, class imbalance remains a problem to tackle. Network overfitting is common in the field of medical imagingbecause of the relatively small size of handled datasets. Data augmentation is classically used in the pre-processing stage to overcome this limitation, and weighted loss functions are designed. For example, Zhao et al. [10,15] used data augmentation by rotating and flipping the heart segmentations to reduce the impact of overfitting. Zhao et al. [14] proposed an automated data augmentation method for synthesizing labeled medical images, which provided significant improvements over state-of-the-art methods for one-shot biomedical image segmentation. Sudre et al. [13] proposed the generalized dice to solve the problem of highly unbalanced segmentations. Abraham et al. [1] proposed a generalized focal loss function based on the Tversky index to address the issue of data imbalance in medical image segmentation. Examples of data augmentation methods to overcome this issue can be found in [2,5–7,12]. However, datasets obtained through data augmentation are strongly correlated with the original datasets, Therefore, the proportion of negative samples remains significantly larger than the proportion of positive samples after data augmentation. Thus, data augmentation does not reduces the risk of overfitting. For the proposed improved loss function can effectively reduce the issues of class imbalance, it does not fundamentally address the problems caused by the lack of datasets.

Therefore, in this paper, in order to segment myocardial pathology (see Fig. 1), we begin with a segmentation of the anatomical tissue (left ventricle (LV), right ventricle (RV), whole heart (WH), myocardium (myo)) around it, and then let the network learn a relationship between these segmentation results to obtain the myocardial pathology. Compared with direct segmentation of myocardial pathology, the effect of class imbalance can be reduced by the segmentation of surrounding anatomical tissues, because it helps the network to focus on the small lesions regarding to the surrounding tissues.

2 Methodology

2.1 Overview of Network Architecture

We propose a hybrid network (see Fig. 2) using 5 UNet [11] to segment myocardial pathology. Our network is composed of three UNet named **UNet1** and two

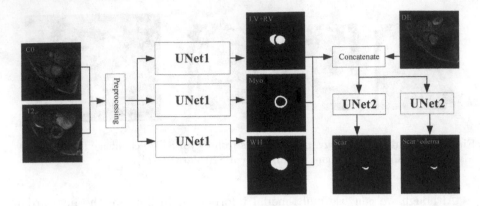

Fig. 2. Global overview of the proposed method.

Table 1. The structural configuration of UNet.

Layers	Input size		Operation	Kernel	Stride	Regul.	Output size	
	UNet1	UNet2					UNet1	UNet2
Input image	(240,240,2)	(240,240,4)	-	-	-	-	(240,240,2)	(240,240,4)
C1	(240,240,2)	(240,240,4)	[Conv2d+relu]*2	3	1	L2	(240,240,64)	(240,240,8)
C2	(240,240,64)	(240,240,8)	Maxpooling2d	2	-	-	(120,120,64)	(120,120,8)
C3	(120,120,64)	(120,120,8)	[Conv2d+relu]*2	3	1	L2	(120,120,128)	(120,120,16)
C4	(120,120,128)	(120,120,16)	Maxpooling2d	2	-	-	(60,60,128)	(60,60,16)
C5	(60,60,128)	(60,60,16)	[Conv2d+relu]*2	3	1	L2	(60,60,256)	(60,60,32)
C6	(60,60,256)	(60,60,32)	Maxpooling2d	2	-	-	(30,30,256)	(30,30,32)
C7	(30,30,256)	(30,30,32)	[Conv2d+relu]*2+Dropout	3	1	L2	(30,30,512)	(30,30,64)
C8	(30,30,512)	(30,30,64)	Maxpooling2d	2	-	-	(15,15,512)	(15,15,64)
C9	(15,15,512)	(15,15,64)	[Conv2d+relu]*2+Dropout	3	1	L2	(15,15,1024)	(15,15,128)
O1	(240,240,2)	(240,240,2)	Sigmoid	-	-	-	(240,240,1)	(240,240,1)

named **UNet2**. The main difference between **UNet1** and **UNet2** is number of filters as shown in Table 1: the number of filters of **UNet1** is [64 128 256 512 256 128 64] and the number of filters of **UNet2** is [8 16 32 64 32 16 8]. Their framework is same. It consists of the classical two parts of the UNet network as shown in Fig. 3: a down-sampling part and an up-sampling part, and shortcut connections between the two parts to fuse high-level features and low-level features. **UNet1** is used to segment the anatomical tissue around myocardial pathology and obtain three segmentation results: LV+RV, Myo, and WH. **UNet2** is used to segment myocardial pathology by learning the relationships between the surrounding anatomical tissue and the pathological ones. Since the lesions are very small and unbalanced, we reduce the number of filters of **UNet2** in order to reduce the impact of overfitting.

3   Experimental Results

Dataset Description. We evaluate our method on the myocardial pathology segmentation combining multi-sequence CMR[1] dataset (MyoPS 2020). Its aim is

---

[1] http://www.sdspeople.fudan.edu.cn/zhuangxiahai/0/MyoPS20/index.html.

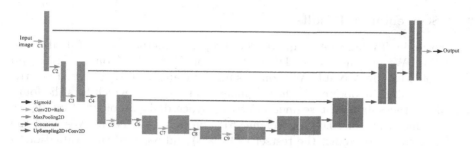

Fig. 3. Architecture of networks.

to segment myocardial pathology, especially scar (infarcted) and edema regions. It contains 45 cases of multi-sequence CMR (25 cases for training and 20 cases for testing). Each case refers to a patient with three sequence CMR, i.e., LGE, T2 and bSSFP CMR. The slice spacings of multi-sequence CMR volume range from 11.999 mm/pixel to 23.000 mm/pixel, while in-plane resolution ranged from 0.729 mm/pixel to 0.762 mm/pixel. The average sizes: $482 \times 479 \times 4$ pixels.

Preprocessing and Postprocessing. We cropped each slice to 240×240 pixels and we do not use data augmentation. The pre-processing begins with a Gaussian normalization. For post-processing, we pad with zeros to get back a initial width and height of a slice.

Implementation and Experimental Setup. We implemented our experiments on Keras/TensorFlow using a NVidia Quadro P6000 GPU. We used five different loss functions for training the network and used sigmoid to get a probability distribution of the left and right ventricle, myocardium, whole heart, scar and edema, and scar, respectively (as shown in Fig. 2). Adam optimizer (batch-size = 1, $\beta_1 = 0.9$, $\beta_2 = 0.999$, $\varepsilon = 0.001$, lr = 0.0001) and did not use learning rate decay. We trained the network during 300 epochs.

Training Step. First, we kept weight of **UNet2** unchanged, which means **UNet2** was not trained at the beginning, then we trained **UNet1**. After finished the train of **UNet1**, we kept weight of **UNet1** unchanged, then trained **UNet2**.

Evaluation Methods. One metric is used to evaluate our method: dice coefficient (DC) to evaluate the regions of myocardial pathology.

3.1 Segmentation Results

As shown in Table 2, we evaluate the proposed method with 5-fold-cross-validation. We obtain a mean DC of 92.3% on WH, 84.9% on LV+RV, and 84.7% on Myo by **UNet1**. Without using data augmentation, based on the original dataset, we obtain a higher segmentation accuracy, which lays the foundation for the subsequent segmentation of myocardial pathology. Finally, we obtain a mean DC of 20.6% on edema, 51% on scar by **UNet2**. We used the trained network to predict the testset (20 cases) and received the evaluation of our prediction results from the MyoPS2020 organizer: the mean DC of 58.6% on scar and the mean DC of 63.9% on scar and edema.

Table 2. Evaluation results on 5-fold-cross-validation.

Patient	101–105	106–110	111–115	116–120	121–125	Average	Test datasets
Edema	0.284	0.153	0.189	0.122	0.280	0.206	–
Scar	0.473	0.496	0.515	0.464	0.602	0.510	0.586
Myo	0.844	0.852	0.811	0.859	0.869	0.847	–
LV+RV	0.818	0.854	0.812	0.897	0.864	0.849	–
WH	0.925	0.937	0.876	0.918	0.959	0.923	–

As shown in Fig. 4, for the segmentation results of whole heart, left and right ventricle, and myocardium, as the number of positive samples continues to decrease, the segmentation accuracy is also decreasing, and false segmentation is mainly concentrated at the boundary, which is mainly because ambiguities often appear near the boundaries of the target domains due to tissue similarities. For the segmentation results of edema and scar, the poorly segmentation result is not only on the boundary, but also in regions. In the original dataset, edema does not exist in many slices, which further leads to a reduction in the effective dataset for edema, therefore, the segmentation network is very difficult to segment edema.

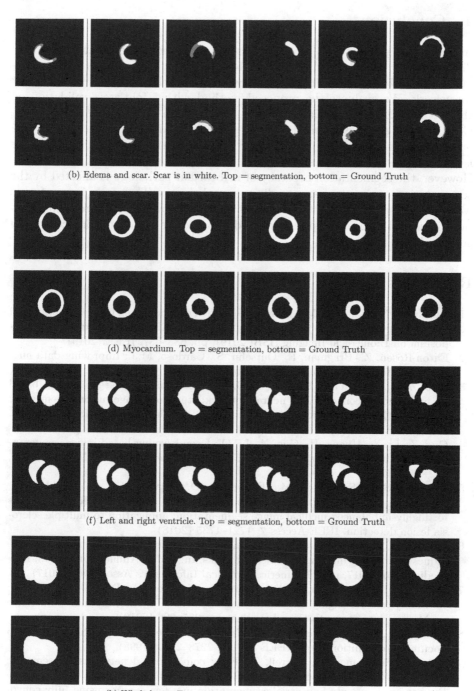

(b) Edema and scar. Scar is in white. Top = segmentation, bottom = Ground Truth

(d) Myocardium. Top = segmentation, bottom = Ground Truth

(f) Left and right ventricle. Top = segmentation, bottom = Ground Truth

(h) Whole heart. Top = segmentation, bottom = Ground Truth

Fig. 4. Qualitative segmentation results.

4   Conclusion

In this paper, we propose a way of reverse thinking, not to segment the myocardial pathology directly, but to learn a relationship between the surrounding normal tissue and it by designing one stacked and parallel UNets with multi-output framework. We evaluate the proposed method with 5-fold-cross-validation on the MICCAI 2020 myocardial pathology segmentation combining multi-sequence CMR Challenge dataset (MyoPS 2020) and achieve a mean DC of 20.6%, 51% on edema and scar, respectively. The computation time of the entire pipeline is less than 3 s for an entire 3D volume, making it usable for clinical practice. However, the segmentation accuracy of myocardial pathology is affected by the segmentation accuracy of surrounding normal tissues. Therefore, in our future work, we will continue to study the relationship between the surrounding normal tissue and myocardial pathology and improve the segmentation accuracy of surrounding normal tissues.

References

1. Abraham, N., Khan, N.M.: A novel focal Tversky loss function with improved attention U-Net for lesion segmentation. In: 2019 IEEE 16th International Symposium on Biomedical Imaging (ISBI 2019), pp. 683–687. IEEE (2019)
2. Eaton-Rosen, Z., Bragman, F., Ourselin, S., Cardoso, M.J.: Improving data augmentation for medical image segmentation (2018)
3. Guo, Z., Li, X., Huang, H., Guo, N., Li, Q.: Medical image segmentation based on multi-modal convolutional neural network: study on image fusion schemes. In: 2018 IEEE 15th International Symposium on Biomedical Imaging (ISBI 2018), pp. 903–907. IEEE (2018)
4. Guo, Z., Li, X., Huang, H., Guo, N., Li, Q.: Deep learning-based image segmentation on multimodal medical imaging. IEEE Trans. Radiat. Plasma Med. Sci. **3**(2), 162–169 (2019)
5. Hashemi, S.R., Salehi, S.S.M., Erdogmus, D., Prabhu, S.P., Warfield, S.K., Gholipour, A.: Asymmetric loss functions and deep densely-connected networks for highly-imbalanced medical image segmentation: application to multiple sclerosis lesion detection. IEEE Access **7**, 1721–1735 (2018)
6. Hussain, Z., Gimenez, F., Yi, D., Rubin, D.: Differential data augmentation techniques for medical imaging classification tasks. In: AMIA Annual Symposium Proceedings, vol. 2017, p. 979. American Medical Informatics Association (2017)
7. Kervadec, H., Bouchtiba, J., Desrosiers, C., Granger, E., Dolz, J., Ayed, I.B.: Boundary loss for highly unbalanced segmentation. In: International Conference on Medical Imaging with Deep Learning, pp. 285–296 (2019)
8. LeCun, Y., Bottou, L., Bengio, Y., Haffner, P.: Gradient-based learning applied to document recognition. Proc. IEEE **86**(11), 2278–2324 (1998)
9. Long, J., Shelhamer, E., Darrell, T.: Fully convolutional networks for semantic segmentation. In: Proceedings of CVPR, pp. 3431–3440 (2015)
10. Puybareau, É., et al.: Left atrial segmentation in a few seconds using fully convolutional network and transfer learning. In: Pop, M., et al. (eds.) STACOM 2018. LNCS, vol. 11395, pp. 339–347. Springer, Cham (2019). https://doi.org/10.1007/978-3-030-12029-0_37

11. Ronneberger, O., Fischer, P., Brox, T.: U-Net: convolutional networks for biomedical image segmentation. In: Navab, N., Hornegger, J., Wells, W.M., Frangi, A.F. (eds.) MICCAI 2015. LNCS, vol. 9351, pp. 234–241. Springer, Cham (2015). https://doi.org/10.1007/978-3-319-24574-4_28

12. Shin, H.-C., et al.: Medical image synthesis for data augmentation and anonymization using generative adversarial networks. In: Gooya, A., Goksel, O., Oguz, I., Burgos, N. (eds.) SASHIMI 2018. LNCS, vol. 11037, pp. 1–11. Springer, Cham (2018). https://doi.org/10.1007/978-3-030-00536-8_1

13. Sudre, C.H., Li, W., Vercauteren, T., Ourselin, S., Jorge Cardoso, M.: Generalised Dice overlap as a deep learning loss function for highly unbalanced segmentations. In: Cardoso, M.J., et al. (eds.) DLMIA/ML-CDS -2017. LNCS, vol. 10553, pp. 240–248. Springer, Cham (2017). https://doi.org/10.1007/978-3-319-67558-9_28

14. Zhao, A., Balakrishnan, G., Durand, F., Guttag, J.V., Dalca, A.V.: Data augmentation using learned transformations for one-shot medical image segmentation. In: Proceedings of the IEEE Conference on Computer Vision and Pattern Recognition, pp. 8543–8553 (2019)

15. Zhao, Z., Boutry, N., Puybareau, É., Géraud, T.: A two-stage temporal-like fully convolutional network framework for left ventricle segmentation and quantification on MR images. In: Pop, M., et al. (eds.) STACOM 2019. LNCS, vol. 12009, pp. 405–413. Springer, Cham (2020). https://doi.org/10.1007/978-3-030-39074-7_42

16. Zhuang, X.: Multivariate mixture model for cardiac segmentation from multi-sequence MRI. In: Ourselin, S., Joskowicz, L., Sabuncu, M.R., Unal, G., Wells, W. (eds.) MICCAI 2016. LNCS, vol. 9901, pp. 581–588. Springer, Cham (2016). https://doi.org/10.1007/978-3-319-46723-8_67

17. Zhuang, X.: Multivariate mixture model for myocardial segmentation combining multi-source images. IEEE Trans. Pattern Anal. Mach. Intell. 41(12), 2933–2946 (2019)

Dual-Path Feature Aggregation Network Combined Multi-layer Fusion for Myocardial Pathology Segmentation with Multi-sequence Cardiac MR

Feiyan Li and Weisheng Li$^{(\boxtimes)}$

Chongqing Key Laboratory of Image Cognition, Chongqing University of Posts
and Telecommunications, Chongqing, China
liws@cqupt.edu.cn

Abstract. Deep convolutional neural networks have shown great potential in medical image segmentation. However, automatic cardiac segmentation is still challenging due to the heterogeneous intensity distributions and indistinct boundaries in cardiac magnetic resonance (CMR) images, especially for myocardial pathology segmentation. In this paper, we present a dual-path feature aggregation network combined multi-layer fusion (MF&DFA-Net) to overcome these misclassification and shape discontinuity problems in myocardial pathology segmentation. The proposed network is aimed to maintain a realistic shape of the segmentation results and predict the position of myocardial pathology, which network is divided into two parts: the first part is a non-downsampling multiscale nested network (MN-Net) which restrains the cardiac shape and maintains the global information, and the second part is multiscale symmetric encoding and decoding network (MSED-Net) that can retain details. Three sequences of CMR images were adopted for multi-layer fusion training which included three inputs and one output. We can segment left ventricular (LV) blood pool, right ventricular (RV) blood pool, LV normal myocardium, LV myocardial edema, LV myocardial scars simultaneously with MF&DFA-Net. We randomly took the 90% data for training and 10% data for verification which data provided by the organizer of the 2020 Medical Image Computing and Computer Assisted Interventions (MIC-CAI) myocardial pathology segmentation challenge (MyoPS 2020). Compared with inter-observer, we increased the Dice value of myocardial scar segmentation by 8.08%.

Keywords: Myocardial pathology segmentation · Dual-path feature aggregation · Multi-layer fusion

1 Introduction

Undoubtedly, medical image segmentation occupies an important position in clinical medicine. Both organ segmentation and lesion segmentation provide doctors with effective auxiliary diagnosis. With the continuous efforts of scientific researchers, semi-automatic and fully automatic segmentation methods for 2D and 3D medical images

© Springer Nature Switzerland AG 2020
X. Zhuang and L. Li (Eds.): MyoPS 2020, LNCS 12554, pp. 146–158, 2020.
https://doi.org/10.1007/978-3-030-65651-5_14

continue to emerge. And with the help of deep neural networks, previous works [1, 2] can quickly and effectively obtain reliable segmentation results.

Cardiac segmentation with high anatomical variability is one of the important tasks in medical image segmentation. Cardiac magnetic resonance imaging (MRI) technology can non-invasively assess heart function. And different sequences of cardiac magnetic resonance (CMR) extract different feature information. There are three different sequences CMR provided by the organizer of the 2020 Medical Image Computing and Computer Assisted Interventions (MICCAI) myocardial pathology segmentation challenge [3, 4]. The bSSFP cine CMR is a balanced steady-state, free precession cine sequence which can learn the cardiac motions and obtain a clear boundary of cardiac. The late gadolinium enhancement (LGE) CMR is a T1-weighted, inversion-recovery, gradient-echo sequence which can enhance the infarcted myocardium, appearing with distinctive brightness compared with the healthy tissues. The T2 CMR is a T2-weighted, black blood spectral presaturation attenuated inversion-recovery sequence which provides imaging of the acute injury and ischemic regions. Please refer to Fig. 1, there shows same slice of three sequences from the same case. It can be clearly seen that the bSSFP cine CMR sequence has stronger contrast than the other two sequences. Therefore, how to effectively and plenty utilize CMR of different sequences has become an urgent problem to be solved.

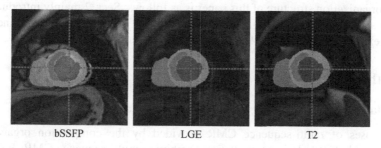

bSSFP   LGE   T2

Fig. 1. Display of the same slice from the same case under different CMR sequences.

With the MICCAI segmentation challenge held year after year, more and more novel multi-sequence CMR segmentation methods have been proposed. For example, in the 2019 multi-sequence CMR segmentation challenge, Chen *et al.* [5] proposed an unsupervised learning method composed of multi-modal image translation network and cascaded segmentation network to segment cardiac structures from LGE CMR without using labelled LGE data for training. Campello *et al.* [6] proposed a new frame composed by shape reconstruction neural network (SRNN) and a spatial constraint network (SCN) to segment LGE CMR. Wang *et al.* [7] adopted an improved U-Net Model with selective kernel to segment multi-sequence CMR. It can be seen that as long as the network structure is properly designed, both supervised learning and unsupervised learning can obtain reliable segmentation results for the same segmentation task. In order to better alleviate the domain shift between different sequences, Wang *et al.* [8] proposed an unsupervised domain alignment method to solve this problem. In addition, there are many excellent solutions in the task of heart organ segmentation. For example,

Li *et al.* [9] proposed multiscale feature aggregation for cardiac right ventricle segmentation. Khened *et al.* [10] proposed fully convolutional multiscale residual DenseNets for cardiac segmentation. In the task of cardiac segmentation, a good level has been achieved. But the intensity inhomogeneity of medical images and the unclear boundary between the region of interest and the background in segmentation tasks which are still challenging. Especially in the myocardial lesion area, the uneven shape of the lesion and the lesion proportion is extremely small, so the segmentation of the myocardial lesion area is still very challenging.

Therefore, this paper proposes a dual-path feature aggregation network combined multi-layer fusion (MF&DFA-Net) to segment the two LV myocardial lesion areas of LV myocardial edema and LV myocardial scars. The network of the first path is a non-downsampling multiscale nested network (MN-Net). The main purpose is to achieve feature reuse through dense connections, and to make up for the missing features of pooling in the second path through non-downsampling. The second path is a multiscale symmetrical encoding and decoding network (MSED-Net), which realizes feature cross-domain cascading and feature reuse through large-scale cross-domain connections and small-scale residual operations. Because the area of the lesion is quite different from the area of the myocardium and the ventricle, different receptive fields are obtained through multiscale convolution in the input part of the two networks to effectively extract the features of different parts.

The organization structure of this paper is as follows: Sect. 2 mainly introduces our methods and data preprocessing. Section 3 shows our experimental results and some ablation experiments, and Sect. 4 summarizes this paper.

2 Methodology

2.1 Data Processing

The 45 cases of multi-sequence CMR provided by the competition organizer of 2020 myocardial pathology segmentation combining multi-sequence CMR. Each case includes bSSFP cine, LGE and T2 CMR sequences. Three sequences are marked by the same set of labels. There are 25 cases used for training and 20 cases for testing. In training data, there are 102 slices and each slice corresponds to a valid label. The provided gold standard labels include: left ventricular (LV) blood pool (labelled 500), right ventricular (RV) blood pool (600), LV normal myocardium (200), LV myocardial edema (1220), LV myocardial scars (2221).

Firstly, we counted the ratio of the effective label pixel values to the entire image pixel values in 102 slices. Please refer to (a) in Fig. 2, in the 83rd slice, the ratio of the pixel value of the marked area to the pixel value of the entire image is the largest. Then we counted the length and width of 25 cases in the training set. As shown in Fig. 2(b), the length and width of the original data are around 500 pixel values. After the length and width of each case are multiplied by the maximum percentage in Fig. 2(a), the length and width are both around 200 pixel values. In order to balance the positive and negative samples of the label but also consider retaining some relevant information, we cropped the center area of training data to [256, 256] like Fig. 3, then we randomly selected 11 slices from training data for verification, the remaining data for training.

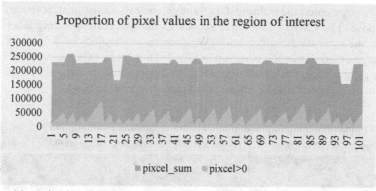

(a)   axis x/y: slice/pixel, max percentage: 91560/230868 ≈ 0.4 (slice 83)

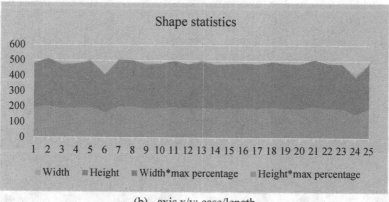

(b)   axis x/y: case/length

Fig. 2. The ratio of the effective label pixel values to the entire image pixel values in 102 slices are shown in (a). The statistics of length and width of 25 cases are shown in (b).

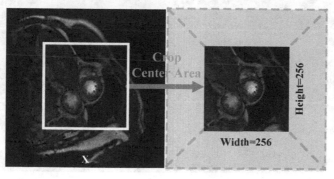

Fig. 3. Schematic diagram of the center area cropping operation.

Last we augmented the training data through contrast adjustment, rotation, flipping and transpose. The CMRs of the three sequences and the corresponding labels had all undergone the same data augmentation. We augmented 91 slices by a factor of 42. The specific augmentation method is shown in Fig. 4. There is an original cropped image in the upper right corner. We conducted rotation, flipping, and transpose operation from top to bottom, and performed contrast adjustment from left to right.

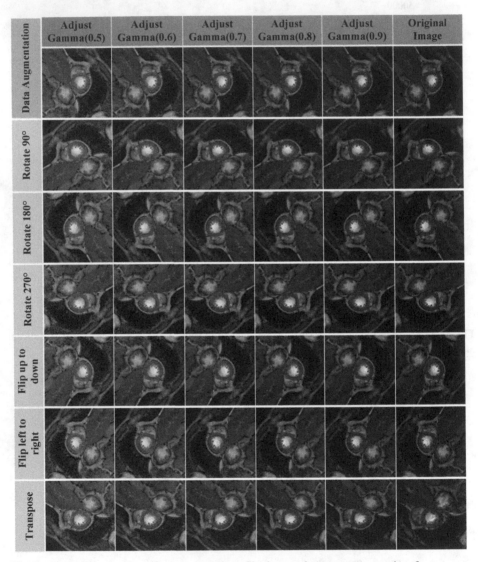

Fig. 4. Data augmentation diagram. Rotation, flipping, and transpose operation from top to bottom, and performed contrast adjustment from left to right.

The contrast adjustment is shown as follows,

$$f(p_i) = \left(\frac{p_i}{p_{max} - p_{min}}\right)^{\gamma} \times (p_{max} - p_{min}), \gamma > 0, i > 0 \qquad (1)$$

where p_i represents i-th pixel value, p_{max} represents max pixel value, p_{min} represents min pixel value. When $\gamma > 1$, the contrast decreases, $\gamma = 1$, the contrast remains unchanged, and $\gamma < 1$, the contrast is increases.

Since the label is composed of multiple classifications, we performed one-hot operation after resetting the pixel value of the label (200 reset to 1, 500 reset to 2, 600 reset to 3, 1220 reset to 4, 2221 reset to 5).

2.2 Proposed Method

The myocardial pathological segmentation solution proposed in this paper is shown in Fig. 5. Three sequences of bSSFP cine, LGE and T2 CMR share encoder 1 and encoder 2. E_1^b, E_1^l, E_1^t represent the encoded output of the bSSFP cine, LGE and T2 CMR sequences in encoder 1, respectively. E_2^b, E_2^l, E_2^t represent the encoded output of the bSSFP cine, LGE and T2 CMR sequences in encoder 2, respectively. $E_1^{b,l,t}$ and $E_2^{b,l,t}$ represent the results of multi-layer fusion, the multi-layer fusion is described in (2) and (3).

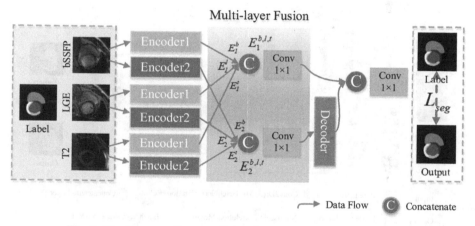

Fig. 5. The diagram of myocardial pathological segmentation solution.

$$E_1^{b,l,t} = \text{concatenate}((E_1^b, E_1^l, E_1^t), axis = -1) \qquad (2)$$

$$E_2^{b,l,t} = \text{concatenate}((E_2^b, E_2^l, E_2^t), axis = -1) \qquad (3)$$

Because the network of the first path is a non-downsampling multiscale nested network and the second path is a multiscale symmetrical encoding and decoding network, we just decoded the output of encoder 2. Finally, we integrate the information of the two paths as the final output. In Fig. 5, the L_{seg} represents loss function for training.

$$L_{dice} = 1 - \frac{2|X^c \cap Y^c| + \varepsilon}{|X^c| + |Y^c| + \varepsilon} \tag{4}$$

$$L_{wce} = -\sum_c w^c X^c \log(Y^c),\, w^c = 1 - p^c / \sum_c p^c \tag{5}$$

$$L_{seg} = L_{dice} + L_{wce} \tag{6}$$

where X^c represents predicted results, Y^c represents ground truth, w^c represents the weight of c th category, p^c represents the sum of pixel values of c th category and c is the category. In experiments, we set $\varepsilon = 1e-5$.

The main details about the encoder and decoder are shown in Fig. 6. The dual-path feature aggregation network combined multi-layer fusion (MF&DFA-Net) is inspired by U-Net [11] and DenseNets [12, 13]. The first path is a non-downsampling MN-Net. The main purpose is to achieve feature reuse through dense connections, and to make up for the missing features of pooling in the second path through non-downsampling. The details are shown in encoder 1. Batch normalization and 'Relu' activation were applied After each 3×3 convolutional layer.

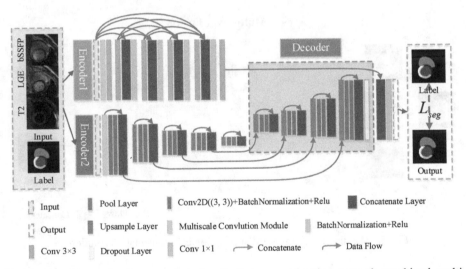

Fig. 6. The Schematic diagram of dual-path feature aggregation network combined multi-layer fusion (MF&DFA-Net). Batch normalization and 'relu' activation are applied after each convolutional layer. A dropout layer (dropout rate = 0.5) is applied to final layer of decoder.

The second path is a MSED-Net, which realizes feature cross-domain cascading and feature reuse through large-scale cross-domain connections and small scale residual operations. In the process of continuous pooling, a lot of information will inevitably be lost. In order to restore the information, the decoding process corresponds to the encoded convolutional layer. The first-path non-downsampling network is complementary to the second-path encoding and decoding network, and finally obtain the segmentation result through cascade.

Because the area of the lesion is quite different from the area of the myocardium and the ventricle, different receptive fields are obtained through multiscale convolution in the input part of the two networks to effectively extract the features of different parts.

The diagram of multiscale convolution module is described as Fig. 7. For the input image, we extracted more features of different receptive fields in the original image through three different scale convolutions, and make the number of channels of each convolutional layer 2 times the number of initial channels (that is, 32 layers), and then, the dimensionality of the feature number is reduced through 1×1 convolution, and finally feature fusion is performed as output.

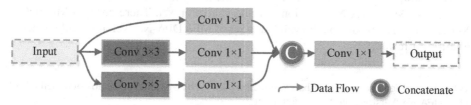

Fig. 7. The diagram of multiscale convolution module (MCM).

3   Experiments and Results

3.1   Implementation Details

Totally, there are 3822 2D slices for training and 11 2D slices for validation. During the training process, we calculated the Dice values of the myocardium, left ventricle, right ventricle, myocardial scars and myocardial edema respectively in the validation set. The calculation of Dice is obtained by

$$Dice^c = \frac{2|X^c \cap Y^c| + \varepsilon}{|X^c| + |Y^c| + \varepsilon}, \varepsilon > 0, c = [0, 5] \tag{7}$$

where c represents the category and $\varepsilon = \text{le}-5$. 2D slices of size 256×256 in pixels are the inputs of MF&DFA-Net. We set the initial learning rate to 0.0001. After 20 iterations of training, when the average Dice values of myocardial scars and myocardial edema in the validation set does not rise anymore, the learning rate is attenuated once, and the attenuation coefficient is 0.1. We use Adam as the optimizer. The experiments were implemented using python and Keras on a NVIDIA Geforce RTX 2080 Ti GPU for training and testing. Due to memory limitation, we set batch size to 4.

The bSSFP, LGE and T2 CMR as inputs of MF&DFA-Net are all normalized as follows,

$$mean = \frac{\sum\limits_{i=1}^{n} X_i}{n} \tag{8}$$

$$stdev = \sqrt{\frac{\sum\limits_{i=1}^{n}(X_i - mean)^2}{n-1}} \tag{9}$$

$$Lambda = \frac{X_i - mean}{stdev + \varepsilon} \tag{10}$$

where ε represents a minimum value that prevents the denominator from being 0 and X_i represents every pixel value of a tensor. We set $\varepsilon = 1e-6$.

3.2 Ablation Experiments and Results

Table 1 shows average Dice score of baseline method, MF&DFA-Net and some ablation methods of MF&DFA-Net Version2 on the validation set. There are two MF&DFA-Net versions. The results on the validation set of MF&DFA-Net Version1 did not adopt multi-layer fusion.

Table 1. Average Dice scores of baseline method, MF&DFA-Net and some ablation methods of MF&DFA-Net Version2 on the validation set.

Method	Validation set average Dice					
	Average	Myocardium	LV	RV	Edema	Scars
U-Net (baseline)	0.6834	0.7706	0.8910	0.8851	0.3207	0.5497
MF&DFA-Net Version1	0.7573	0.8340	**0.9366**	0.9117	0.3930	0.7112
MF&DFA-Net Version2	**0.8172**	**0.8480**	0.9169	**0.9311**	**0.6306**	**0.7596**
Version2 w/o MCM	0.6917	0.7739	0.8921	0.8920	0.3228	0.5778
Version2 w/o Encoder1	0.6862	0.7768	0.8915	0.8941	0.3068	0.5622
Version2 w/o MCM & Encoder1	0.6848	0.7821	0.8900	0.8915	0.2999	0.5605

Table 1 also shows the results of MF&DFA-Net Version2 without MCM and Encoder simultaneously and separately on the validation set. In the ablation experiments, the experimental results without Encoder1 and MCM are equivalent to U-Net. The results on the validation set of MF&DFA-Net Version1 achieved best LV segmentation results. The results on the validation set of MF&DFA-Net Version2 achieved best segmentation results except LV, especially for myocardial edema and myocardial scars.

There are 20 cases in test set, we take one slice as an example to show the visualization results on bSSFP cine, LGE and T2 CMR sequences as shown in Fig. 8. There is no labels in test set, we can only roughly analyze the quality of the results through the continuity and completeness of the test results. But the visualization results of MF&DFA-Net Version1 and MF&DFA-Net Version2 are consistent in test results.

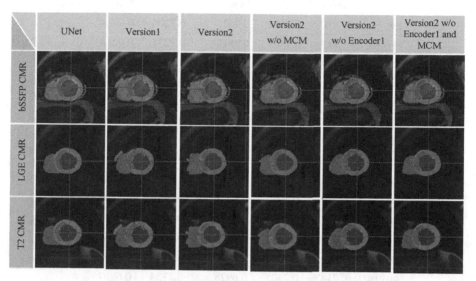

Fig. 8. The visualization test results under different methods on test set.

3.3 Test Results

Table 2 shows the submitted results of MF&DFA-Net Version1 and MF&DFA-Net Version2 on the test set. The MF&DFA-Net Version1 did not adopt multi-layer fusion and fine tune well. In other words, Version1 adopted multiscale dual-path feature aggregation network to train the data of the three sequences separately, and then takes the best test result of the three sequences.

After conducting multi-layer fusion, the Dice score has been greatly improved. The bSSFP cine CMR is a balanced steady-state, free precession cine sequence which can learn the cardiac motions and obtain a clear boundary of cardiac. The late gadolinium enhancement (LGE) CMR is a T1-weighted, inversion-recovery, gradient-echo sequence which can enhance the infarcted myocardium, appearing with distinctive brightness compared with the healthy tissues. The T2 CMR is a T2-weighted, black blood spectral presaturation attenuated inversion-recovery sequence which provides imaging of the acute injury and ischemic regions. So the multi-level fusion effectively extracted and integrated the features of the three different CMR sequences.

Table 2. Dice scores of MF&DFA-Net Version1 and MF&DFA-Net Version2 on the test set.

Results submitted	Test set average Dice			
	Version1 Dice		Version2 Dice	
Case	Scar	Edema + Scar	Scar	Edema + Scar
'myops_test_201'	0.7621	0.7621	0.7200	0.7384
'myops_test_202'	0.2306	0.3548	0.2271	0.4214
'myops_test_203'	0.5878	0.5551	0.5819	0.5225
'myops_test_204'	0.5019	0.5576	0.7725	0.7561
'myops_test_205'	0.5254	0.5859	0.6472	0.6394
'myops_test_206'	0.7292	0.6169	0.7748	0.7608
'myops_test_207'	0.0000	0.5194	0.0000	0.5470
'myops_test_208'	0.6160	0.6196	0.6555	0.7073
'myops_test_209'	0.6666	0.7164	0.7412	0.7069
'myops_test_210'	0.4092	0.5005	0.0626	0.5701
'myops_test_211'	0.5863	0.5773	0.7677	0.7062
'myops_test_212'	0.7478	0.7075	0.7625	0.7761
'myops_test_213'	0.2059	0.2865	0.1958	0.2798
'myops_test_214'	0.7405	0.6928	0.7754	0.6997
'myops_test_215'	0.7671	0.7697	0.8543	0.7543
'myops_test_216'	0.6699	0.6875	0.6398	0.6299
'myops_test_217'	0.7818	0.8118	0.8272	0.7955
'myops_test_218'	0.1692	0.2500	0.5577	0.5410
'myops_test_219'	0.4167	0.4876	0.7504	0.7756
'myops_test_220'	0.6397	0.6206	0.7876	0.7849
mean	0.5377	0.5840	**0.6051**	**0.6557**
std	0.2296	0.1542	**0.2630**	**0.1376**

There is one case obtained 0 for the Dice score which is special and difficult. Dice score for myocardial scar reaches 0.6051 ± 0.2630. The inter-observer variation of manual myocardial scar segmentation, in terms of Dice overlap, was 0.5243 ± 0.1578. Partial results of MF&DFA-Net Version2 shown in Fig. 9.

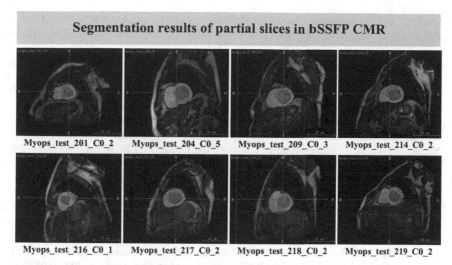

Fig. 9. Partial visualization test results of MF&DFA-Net Version2 on test set.

4 Conclusion

In this paper, we propose a dual-path feature aggregation network combined multi-layer fusion (MF&DFA-Net) to overcome these misclassification and shape discontinuity problems in multi-sequence myocardial pathology segmentation. We trained our model on 3822 slices and verified on 11 slices. We further verified the reliability of proposed method through ablation experiments, especially the effect of multi-layer fusion on multi-sequence myocardial pathological segmentation. We tested on 20 cases, the Dice value of myocardial scars in our submitted test results exceeds inter-observer by 8.08%.

Acknowledgements. This work was supported by the National Natural Science Foundation of China [Nos. 61972060 and U1713213], National Science & Technology Major Project [2016YFC1000307-3], Natural Science Foundation of Chongqing [cstc2019cxcyljrc-td0270, cstc2019jcyj-cxttX0002, cstc2019jcyj-zdxmX0011].

References

1. Petitjean, C., Dacher, J.: A review of segmentation methods in short axis cardiac MR images. Med. Image Anal. **15**(2), 169–184 (2011)
2. Oktay, O., et al.: Anatomically constrained neural networks (ACNNs): application to cardiac image enhancement and segmentation. IEEE Trans. Med. Imaging **37**(2), 384–395 (2018)
3. Zhuang, X.: Multivariate mixture model for myocardial segmentation combining multi-source images. IEEE Trans. Pattern Anal. Mach. Intell. **41**(12), 2933–2946 (2019)
4. Zhuang, X.: Multivariate mixture model for cardiac segmentation from multi-sequence MRI. In: Ourselin, S., Joskowicz, L., Sabuncu, M.R., Unal, G., Wells, W. (eds.) MICCAI 2016. LNCS, vol. 9901, pp. 581–588. Springer, Cham (2016). https://doi.org/10.1007/978-3-319-46723-8_67

5. Chen, C., et al.: Unsupervised multi-modal style transfer for cardiac MR segmentation. In: Pop, M., et al. (eds.) STACOM 2019. LNCS, vol. 12009, pp. 209–219. Springer, Cham (2020). https://doi.org/10.1007/978-3-030-39074-7_22

6. Campello, V.M., Martín-Isla, C., Izquierdo, C., Petersen, S.E., Ballester, M.A.G., Lekadir, K.: Combining multi-sequence and synthetic images for improved segmentation of late gadolinium enhancement cardiac MRI. In: Pop, M., et al. (eds.) STACOM 2019. LNCS, vol. 12009, pp. 290–299. Springer, Cham (2020). https://doi.org/10.1007/978-3-030-39074-7_31

7. Wang, X., et al.: SK-Unet: an improved U-Net model with selective kernel for the segmentation of multi-sequence cardiac MR. In: Pop, M., Sermesant, M., Camara, O., Zhuang, X., Li, S., Young, A., Mansi, T., Suinesiaputra, A. (eds.) STACOM 2019. LNCS, vol. 12009, pp. 246–253. Springer, Cham (2020). https://doi.org/10.1007/978-3-030-39074-7_26

8. Wang, J., Huang, H., Chen, C., Ma, W., Huang, Y., Ding, X.: Multi-sequence cardiac MR segmentation with adversarial domain adaptation network. In: Pop, M., Sermesant, M., Camara, O., Zhuang, X., Li, S., Young, A., Mansi, T., Suinesiaputra, A. (eds.) STACOM 2019. LNCS, vol. 12009, pp. 254–262. Springer, Cham (2020). https://doi.org/10.1007/978-3-030-39074-7_27

9. Li, J., Yu, Z., Gu, Z., Liu, H., Li, Y.: Dilated-Inception Net: multi-Scale feature aggregation for cardiac right ventricle segmentation. IEEE Trans. Bio-Med. Eng. 66(12), 3499–3508 (2019)

10. Khened, M., Alex, V., Krishnamurthi, G.: Fully convolutional multi-scale residual DenseNets for cardiac segmentation and automated cardiac diagnosis using ensemble of classifiers. Med. Image Anal. 51, 21–45 (2018)

11. Ronneberger, O., Fischer, P., Brox, T.: U-Net: convolutional networks for biomedical image segmentation. In: Navab, N., Hornegger, J., Wells, W.M., Frangi, A.F. (eds.) MICCAI 2015. LNCS, vol. 9351, pp. 234–241. Springer, Cham (2015). https://doi.org/10.1007/978-3-319-24574-4_28

12. Jégou, S., Drozdzal, M., Vázquez, D., Romero, A., Bengio, Y.: The one hundred layers tiramisu: fully convolutional DenseNets for semantic segmentation. In: Proceedings of the IEEE Conference on Computer Vision and Pattern Recognition, pp. 1175–1183 (2017)

13. Huang, G., Liu, Z., Van Der Maaten, L., Weinberger, K.Q.: Densely connected convolutional networks. In: Proceedings of the IEEE Conference on Computer Vision and Pattern Recognition, pp. 4700–4708 (2017)

Cascaded Framework with Complementary CMR Information for Myocardial Pathology Segmentation

Jun Ma(✉)(iD)

Department of Mathematics, Nanjing University of Science and Technology,
Nanjing, China
junma@njust.edu.cn

Abstract. Myocardial pathology segmentation in cardiac magnetic resonance (CMR) is an important step for patients suffering from myocardial infarction. In this paper, we present a cascaded framework with complementary information for infarcted and edema regions segmentation in CMR sequences. Specifically, instead of using all the three CMR sequences as joint inputs, we first use a 2D U-Net with balanced-Steady State Free Precession (bSSFP) cine sequence to segment the whole heart (left ventricle and myocardium) because bSSFP can capture cardiac motions and present clear boundaries. Then, we crop the whole heart as a region of interest (ROI). Finally, we segment the scar and edema regions in the late gadolinium enhancement (LGE) and T2 CMR sequence ROI. We evaluate the proposed method on MICCAI 2020 MyoPS testing set and achieve Dice scores 0.6283 ± 0.2772 for scar and 0.5419 ± 0.2406 for the combination of edema and scar, which is better than the inter-observer variation of manual scar segmentation (0.5243 ± 0.1578).

Keywords: Segmentation · Myocardial pathology · Cascaded framework

1 Introduction

Quantitative assessment of myocardial viability is essential in the diagnosis and treatment management for patients suffering from myocardial infarction (MI). Cardiac magnetic resonance (CMR) is particularly used to provide imaging anatomical and functional information of heart, such as the late gadolinium enhancement (LGE) CMR sequence which visualizes MI, the T2-weighted CMR which images the acute injury and ischemic regions, and the balancedSteady State Free Precession (bSSFP) cine sequence which captures cardiac motions and presents clear boundaries. Combining these multi-sequence CMR data can provide rich and reliable information as well as morphological information of the myocardium [9].

One of the important tasks is to segment the myocardium into different regions, including normal myocardium, infarction and edema, from multi-sequence CMR dataset. Manual annotation is generally time-consuming, tedious

© Springer Nature Switzerland AG 2020
X. Zhuang and L. Li (Eds.): MyoPS 2020, LNCS 12554, pp. 159–166, 2020.
https://doi.org/10.1007/978-3-030-65651-5_15

and subjects to inter- and intra-observer variations. Thus, fully automatic segmentation method is highly desired in clinical practice. Figure 1 presents some images from different CMR sequences and the corresponding edema and scars annotations. It can be observed that the intensity appearances vary significantly among different sequences, and the both edema and scars have ambiguous boundaries and low contrast. Thus, it is very challenging to automatically segment them.

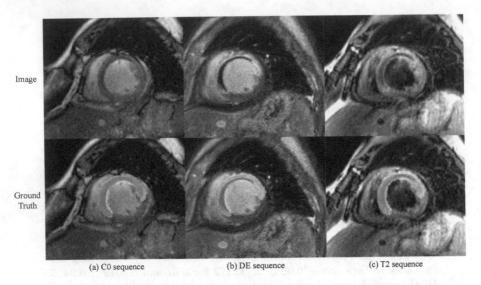

Image

Ground
Truth

(a) C0 sequence   (b) DE sequence   (c) T2 sequence

Fig. 1. Visual examples of different CMR sequence images. C0, DE, and T2 stand for the balanced-Steady State Free Precession (bSSFP) cine sequence, the late gadolinium enhancement (LGE) CMR sequence, and the T2-weighted CMR, respectively. In the second row, the gray and light green color denote myocardial edema and myocardial scar respectively. (Color figure online)

To the best of our knowledge, most of CMR segmentation related studies focus on left ventricle, right ventricle, and myocardium segmentation [1,2,10], little work has been done in the fully automatic cardiac pathology segmentation [4,5,8]. Zhuang [8] proposed a multivariate mixture model and maximum of log-likelihood framework for simultaneous registration and segmentation of multi-source CMR images, achieving a Dice score of 0.4779 ± 0.1855 for scars segmentation. Recently, Li et al. [5] proposed a new framework of scar quantification based on surface projection and graph-cuts framework, achieving a mean accuracy of 0.856 ± 0.033 and mean Dice score of 0.702 ± 0.071 for LA scar quantification.

2   Method

This paper focuses on myocardial scar and edema segmentation from the following three CMR sequences

- C0 sequence; It is a balanced steady-state, free precession (bSSFP) cine sequence, which captures cardiac motions and presents clear boundaries;
- DE sequence; It is late gadolinium enhancement (LGE) CMR sequence, which visualizes myocardial infarction (MI);
- T2 sequence; It is T2-weighted CMR, which visualizes acute injury and ischemic regions.

One of the main challenges is how to combine these multi-sequence CMR data and exploit rich and reliable information regarding to the pathological as well as morphological information of the myocardium.

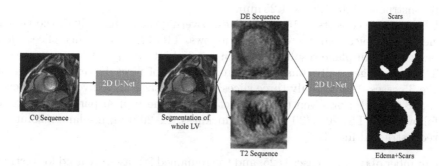

Fig. 2. Pipeline of the proposed method. Due to C0 sequence can present clear boundaries of left ventricular (LV), we first use a 2D U-Net to segment whole LV from the C0 sequence, including LV blood pool and myocardium. Then, we crop the LV region of interest (ROI) from DE sequence and T2 sequence. The pathology is relatively more clear in DE and T2. Thus, a new 2D U-Net is used to segment the scars and the combination of scars and edema from the DE sequence and T2 sequence.

Motivated by the characteristics of different CMR sequences, we propose a cascaded framework for myocardial edema and scar segmentation, which can exploit the complementary information of the three CMR sequences. Figure 2 presents the whole pipeline of the proposed method. Specifically, the proposed method contains three steps[1]:

- Step 1 (whole LV segmentation). Train a 2D U-Net [6] to segment the whole LV (including left ventricular blood pool and myocardium) from C0 sequence, because the heart boundary is clear in this sequence;
- Step 2 (creating ROI). Crop LV region of interest (ROI) from DE and T2 sequence based on the segmentation results in step 1. In this way, the unrelated background can be excluded;
- Step 3 (scar and edema segmentation). Train a new 2D U-Net to segment the scar and edema from DE and T2 sequences because the pathology is more clear in the two sequences. Specifically, DE and T2 sequences are combined as two channels and then input to the network.

---

[1] In step 1 and step 3, the networks are trained end-to-end, while the whole framework is not end-to-end.

3   Experiments and Results

3.1   Dataset and Training Protocols

Dataset. Three-sequence CMR from 45 patients [7, 8] are involved in this study. Specifically,

- C0 sequence generally consists of 8 to 12 contiguous slices, covering the full ventricles from the apex to the basal plane of the mitral valve, with some cases having several slices beyond the ventricles. The typical parameters are as follows, TR/TE: 2.7/1.4 ms; slice thickness: 8–13 mm; inplane resolution: reconstructed into 1.25 × 1.25 mm.
- DE sequence consists of 10 to 18 slices, covering the main body of the ventricles. The typical parameters are as follows, TR/TE: 3.6/1.8 ms; slice thickness: 5 mm; in-plane resolution: reconstructed into 0.75 × 0.75 mm.
- T2 sequence generally consists of a small number of slices. For example, among the 35 cases, 13 have only three slices, and the others have five (13 subjects), six (8 subjects) or seven (one subject) slices. The typical parameters are as follows, TR/TE: 2000/90 ms; slice thickness: 12–20 mm; in-plane resolution: reconstructed into 1.35 × 1.35 mm.

The number of training cases is 25, and the remained 20 cases are used for testing. During preprocessing, we apply z-score to separately normalize each sequence.

We employ nnU-Net [3] as the main network. Due to the fact that the CMR data has large slice thickness, 2D U-Net is more suitable in this task. During training, the patch size is 112 × 112 and batch size is 6. We apply five-fold cross validation in all experiments. Each fold is trained on a TITAN V100 GPU.

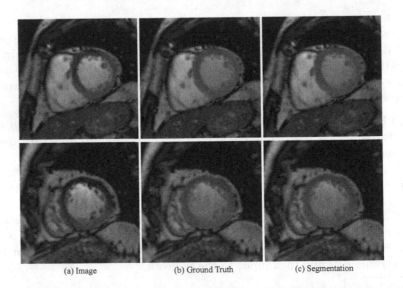

    (a) Image   (b) Ground Truth   (c) Segmentation

Fig. 3. Visual examples of the whole LV segmentation results.

Table 1. Five-fold cross validation results of the whole LV segmentation.

Fold	0	1	2	3	4	Average
Dice	0.9651	0.9613	0.9558	0.9659	0.9665	0.9629

3.2 Five-Fold Cross Validation Results of the Whole LV Segmentation

Table 1 shows five-fold cross validation results for the whole LV segmentation, and Fig. 3 presents some examples of the segmentation results. It can be found that the segmentation results are quite accurate, where the average Dice score in each fold is more than 0.95. The high LV segmentation accuracy can insure that all the myocardial lesions (scar and edema) are included in the segmentation ROI. Thus, when we crop the LV ROI from DE and T2 CMR sequences based on the segmentation results.

3.3 Five-Fold Cross Validation Results of the Pathology Segmentation

Table 2 shows the five-fold cross validation results of scar and edema segmentation. We conduct two groups of experiments: only using DE CMR sequence and using both DE and T2 sequence. Results show that using two sequences can obtain better performance, especially for Edema + Scar, with up to 10% improvements in terms of Dice.

Table 2. Five-fold cross validation results of scar and edema segmentation based on only DE sequence and both DE and T2 sequence, respectively.

Sequence	Fold	Scar Dice	Edema + Scar Dice
DE	0	0.5608	0.5372
	1	0.6336	0.6049
	2	0.5176	0.4659
	3	0.621	0.6332
	4	0.4675	0.4995
	Average	0.5601	0.54814
DE+T2	0	0.5626	0.6512
	1	0.6864	0.6925
	2	0.5199	0.5847
	3	0.6241	0.6931
	4	0.4912	0.6522
	Average	0.57684	0.65474

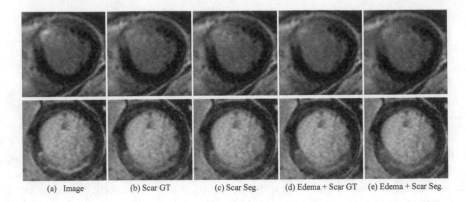

(a) Image   (b) Scar GT   (c) Scar Seg.   (d) Edema + Scar GT   (e) Edema + Scar Seg.

Fig. 4. Visual examples of the scar and edema segmentation results from validation set.

Figure 4 presents some examples of the scar and edema segmentation results. The boundaries of edema and scar are very unclear as show in Fig. 4-(a), which are extremely challenging. There are obvious errors in the segmentation results, which is in accordance with the relatively low Dice scores in Table 2.

3.4 Pathology Segmentation Results on Testing Set

Table 3 shows the quantitative segmentation results for each case in testing set. Some cases (e.g., myops_2204, myops_2215) obtain good segmentation performance for scar segmentation, with 0.8+ in Dice. However, some cases (e.g.., myops_2207, myops_2218) are failed with almost zero Dice. Figure 5 presents the box plots to visualize the quantitative results. It should be noted that the Dice of Edema + Scar is significantly worse than the Dice of Scar, indicating that the segmentation results of edema is much more worse than scar. Figure 6 presents some visualized segmentation results of edema and scar.

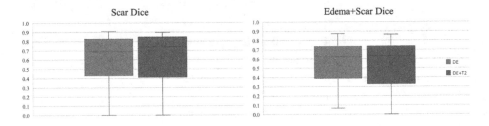

Fig. 5. Box plots of testing set segmentation results.

Table 3. Quantitative scar and edema segmentation results on testing set.

Cases	DE		DE+T2	
	Scar Dice	Edema + Scar Dice	Scar Dice	Edema + Scar Dice
myops_2201	0.6468	0.5455	0.5580	0.4367
myops_2202	0.1721	0.4020	0.0949	0.2583
myops_2203	0.5212	0.4981	0.5086	0.3762
myops_2204	0.8446	0.6497	0.7453	0.5704
myops_2205	0.6829	0.6616	0.7479	0.6660
myops_2206	0.7602	0.7650	0.8490	0.7861
myops_2207	0.0000	0.1789	0.0000	0.0000
myops_2208	0.7796	0.6970	0.7148	0.6631
myops_2209	0.6947	0.5995	0.8222	0.6716
myops_2210	0.2754	0.0667	0.2574	0.1453
myops_2211	0.8289	0.7182	0.8583	0.7013
myops_2212	0.8307	0.6499	0.8962	0.6610
myops_2213	0.4314	0.3867	0.2912	0.2681
myops_2214	0.4294	0.3171	0.7333	0.5605
myops_2215	0.9076	0.8730	0.8938	0.8652
myops_2216	0.5432	0.4689	0.6848	0.6075
myops_2217	0.8107	0.7558	0.8327	0.7463
myops_2218	0.1593	0.1478	0.3782	0.3135
myops_2219	0.8289	0.8178	0.7876	0.7820
myops_2220	0.7517	0.7389	0.8516	0.7587
Average	0.5950	0.5469	0.6253	0.5419
Standard deviation	0.2680	0.2328	0.2772	0.2406

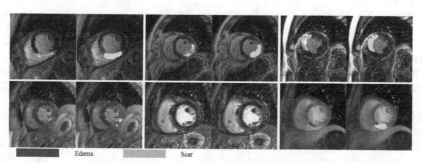

Fig. 6. Visual examples of the scar and edema segmentation results on testing set.

4 Conclusion

Myocardial pathology segmentation is a challenging task due to its unclear boundaries and low contrast in CMR sequences. In this paper, we designed a cascaded framework that enables to utilize the complementary informations in different CMR sequences. Experiments on MICCAI 2020 MyoPS testing dataset show that the proposed method can achieve better performance than the inter-observer variation.

Acknowledgement. This work is supported by the National Natural Science Foundation of China (No. 91630311, No.11971229). The author highly appreciates the organizers of Myocardial pathology segmentation combining multi-sequence CMR (MyoPS 2020) for their public dataset and organizing the great challenge.

References

1. Bernard, O., et al.: Deep learning techniques for automatic MRI cardiac multi-structures segmentation and diagnosis: is the problem solved? IEEE Trans. Med. Imaging **37**(11), 2514–2525 (2018)
2. Chen, C., et al.: Deep learning for cardiac image segmentation: a review. Front. Cardiovas. Med. **7**, 25 (2020)
3. Isensee, F., Jäger, P.F., Kohl, S.A., Petersen, J., Maier-Hein, K.H.: Automated design of deep learning methods for biomedical image segmentation. arXiv preprint arXiv:1904.08128 (2020)
4. Li, L., Weng, X., Schnabel, J.A., Zhuang, X.: Joint left atrial segmentation and scar quantification based on a dnn with spatial encoding and shape attention. arXiv preprint arXiv:2006.13011 (2020)
5. Li, L., et al.: Atrial scar quantification via multi-scale CNN in the graph-cuts framework. Med. Image Anal. **60**, 101595 (2020)
6. Ronneberger, O., Fischer, P., Brox, T.: U-net: convolutional networks for biomedical image segmentation. In: Navab, N., Hornegger, J., Wells, W.M., Frangi, A.F. (eds.) Medical Image Computing and Computer-Assisted Intervention, pp. 234–241 (2015)
7. Zhuang, X.: Multivariate mixture model for cardiac segmentation from multi-sequence MRI. In: Ourselin, S., Joskowicz, L., Sabuncu, M.R., Unal, G., Wells, W. (eds.) MICCAI 2016, Part II. LNCS, vol. 9901, pp. 581–588. Springer, Cham (2016). https://doi.org/10.1007/978-3-319-46723-8_67
8. Zhuang, X.: Multivariate mixture model for myocardial segmentation combining multi-source images. IEEE Trans. Pattern Anal. Mach. Intell. **41**(12), 2933–2946 (2019)
9. Zhuang, X., Li, L.: Multi-sequence CMR based myocardial pathology segmentation challenge (2020). https://doi.org/10.5281/zenodo.3715932
10. Zhuang, X., et al.: Evaluation of algorithms for multi-modality whole heart segmentation: an open-access grand challenge. Med. Image Anal. **58**, 101537 (2019)

Recognition and Standardization of Cardiac MRI Orientation via Multi-tasking Learning and Deep Neural Networks

Ke Zhang and Xiahai Zhuang$^{(\boxtimes)}$

School of Data Science, Fudan University, Shanghai, China
{16307100128,zxh}@fudan.edu.cn

Abstract. In this paper, we study the problem of imaging orientation in cardiac MRI, and propose a framework to categorize the orientation for recognition and standardization via deep neural networks. The method uses a new multi-tasking strategy, where both the tasks of cardiac segmentation and orientation recognition are simultaneously achieved. For multiple sequences and modalities of MRI, we propose a transfer learning strategy, which adapts our proposed model from a single modality to multiple modalities. We embed the orientation recognition network in a Cardiac MRI Orientation Adjust Tool, i.e., CMRadjustNet. We implemented two versions of CMRadjustNet, including a user-interface (UI) software, and a command-line tool. The former version supports MRI image visualization, orientation prediction, adjustment, and storage operations; and the latter version enables the batch operations. The source code, neural network models and tools have been released and open via https://zmiclab.github.io/projects.html.

Keywords: Orientation recognition · Multi-task learning · Cardiac MRI

1 Introduction

Cardiac Magnetic Resonance (CMR) images could be stored in different image orientations when they are recorded in DICOM format and stored into the PACS systems. Recognizing and understanding this difference is crucial in deep neural network (DNN)-based image processing and computing, since current DNN systems generally only take the input and output of images as matrices or tensors, without considering the imaging orientation and real world coordinate. This work is aimed to provide a study of the CMR image orientation, for reference to the

X. Zhuang—This work was funded by the National Natural Science Foundation of China (Grant No. 61971142), and Shanghai Municipal Science and Technology Major Project (Grant No. 2017SHZDZX01).

© Springer Nature Switzerland AG 2020
X. Zhuang and L. Li (Eds.): MyoPS 2020, LNCS 12554, pp. 167–176, 2020.
https://doi.org/10.1007/978-3-030-65651-5_16

human anatomy and standardized coordinate system of real world, and to develop an efficient method for recognition and standardization of the orientation.

Deep neural networks have been demonstrated to achieve state-of-art performance in many medical imaging tasks, such as image segmentation and lesion detection. For CMR images, standardization of all the images is a prerequisite for further computing tasks based on DNN-based methodologies, such as image segmentation [2,5,13] and myocardial pathology analysis [12].

Nevertheless, recognizing the orientation of different modality CMR images and adjusting them into standard format could be as challenging as the further computing tasks. Different from other work that focuses on segmentation or classification individually [10] or combine image segmentation with quantification [6] this work proposes a DNN-based framework to solve the cardiac image segmentation and orientation recognition tasks simultaneously.

The original multi-tasking learning aims at exploiting commonalities and differences across tasks. To extend this concept to deep learning, the multi-tasking framework trains the neural network to learn from different tasks and solve different medical image processing tasks at the same time. In recent years, multi-tasking methods in medical image processing have grown in popularity. Xue et al. propose a multitask learning network (FullLVNet) [9], which modeled intra- and inter-task relatedness to enforce improvement of generalization. Vigneault et al. presented the Ω-Net (Omega-Net): a convolutional neural network (CNN) architecture for simultaneous localization, transformation into a canonical orientation, and semantic segmentation [7]. The auxiliary task could improve the generalization performance by concurrently learning with the main task, which is the main merit if multi-task learning [4]. The previous work focuses on several segmentation tasks or deal with segmentation and classification tasks at the same time. To tackle the difference of CMR image orientation when they were presented for DNN-based image processing, we propose a recognition task, and combine it with the traditional task of CMR image segmentation.

Deep learning-based methods have been widely used in orientation recognition and prediction tasks. Wolterink et al. proposed an algorithm that extracts coronary artery centerlines in cardiac CT angiography (CCTA) images using a convolutional neural network (CNN) [8]. Duan et al. combine a multi-task deep learning approach with atlas propagation to develop a shape-refined biventricular segmentation pipeline for short-axis CMR volumetric images [3]. Based on CMR orientation recognition, we further develop a framework for standardization and adjustment of the orientation.

Given that many clinical applications rely on both an accurate segmentation and orientation recognition to extract specific anatomy or compute some functional indices, we therefore further propose a new multi-task learning framework that aims to solve the CMR orientation recognition and cardiac segmentation tasks at the same time. To enable the proposed model to be conveniently applied in medical image processing and clinical practice, we develop a CMR Orientation Adjust Tool, resulting in a DNN model referred to as CMRadjustNet. For simplicity, the CMR Orientation Adjust Tool is referred to as the CMRadjustNet Tool or CMRadjustNet in the remaining of the article. CMRadjustNet is

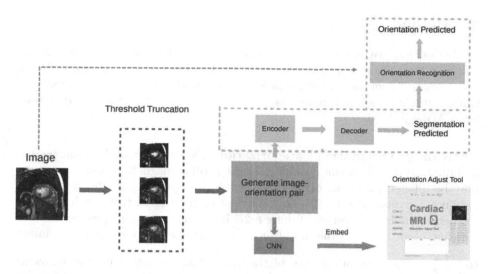

Fig. 1. The pipeline of the proposed CMR orientation recognition and standardization method. The image is first truncated at several gray value thresholds. Then the processed image is used to generate the image-orientation pair (see Sect. 2.1). Then, the multi-tasking network generates the orientation predicted and segmentation mask. When embed the orientation recognition network to the orientation adjust tool, the multi-tasking network is replaced with a simplified CNN.

embedded with the proposed simplified orientation recognition network and support orientation recognition and standardization automatically. The tool has two versions, a graphical interface, and a command-line tool. The graphics interface version supports the display of MRI slices and orientation recognition, standardization, save functions. We developed a user-friendly graphical interface to help users perform CMR image orientation standardization easily. The command-line tool supports batch orientation correction of all MRI files in a folder, which facilitates the processing of large amounts of MRI data.

This work is aimed at designing a DNN-based approach to achieve orientation recognition and standardization for multiple CMR modalities. Figure 1 presents the pipeline of our proposed method. The main contributions of this work are summarized as follows:

(1) We propose a scheme to standardize the CMR image orientation and categorize all the orientations for classification.
(2) We present a DNN-based orientation recognition method for CMR image and transfer it to other modalities.
(3) We propose a multi-tasking network, where orientation recognition and image segmentation tasks are implemented simultaneously.
(4) We develop a CMR image orientation adjust tool (CMRadjustNet) embedded with a simplified orientation recognition network, which facilitates the CMR image orientation recognition and standardization in clinical and medical image processing practice.

2   Method

In this section, we introduce our proposed method for orientation recognition and standardization. Our proposed framework is built on the categorization of CMR image orientations. We propose a multi-tasking network and embed the simplified orientation recognition network into the CMR orientation adjust tool, i.e., CMRadjustNet.

CMR Image Orientation Categorization. Due to different data sources and scanning habits, the orientation of different cardiac magnetic resonance images may be different, and the orientation vector corresponding to the image itself may not correspond correctly. This may cause problems in tasks such as image segmentation or registration. Taking a 2D image as an example, we set the orientation of an image as the initial image and set the four corners of the image to $\begin{array}{cc} 1 & 2 \\ 3 & 4 \end{array}$, Then the orientation of the 2D MR image may have the following 8 variations, which is listed in Table 1. For each image label pair X_t, Y_t. One target orientation O_t is randomly picked from the 9 orientation classes, correspondingly, we flip X_t, Y_t towards the picked orientation. Then we obtain the image-label pair X_t, Y_t and image-orientation pair X_t, O_t . We denote the process of generating image-orientation pair as function g.

Multi-tasking Network. Suppose given image-label pair (X_t, Y_t), X_t is then normalized. We denote the processed X_t as X'. After generating image-orientation pair, $g(X')$ is taken as the input of multi-tasking network. Denote the encoder of proposed multi-tasking network as *encoder*, the decoder of proposed multi-tasking network as *decoder*, the model pipeline is formulated as below:

$$X_{\text{feature}} = encoder(g(X'))$$

$$\hat{Y} = decoder(X_{\text{feature}})$$

$$\hat{O} = F_{\text{orientation}}([g(X'), \hat{Y}]).$$

Here, $\hat{Y}$ and $g(X')$ are concatenated and pass through orientation recognition sub-network $F_{\text{orientation}}$ to predict orientation classification. We start with the orientation recognition branch in the multi-tasking network, where we are interested in take segmentation masks predicted by segmentation network as attention map. In the proposed multi-tasking framework, the orientation recognition sub-network consists of 3 convolution layer and a fully connected layers. The orientation predicted is denoted as $\hat{O}$. We use the standard categorical loss to calculate the loss between predicted orientation $\hat{O}$ and orientation label O,

$$L_{orientation} = \sum_{i=1}^{C} O_i log(\hat{O}_i).$$

Table 1. Orientation categorization of 2D CMR Images. Here, sx, sy and sz respectively denote the size of image in X-axis, Y-axis and Z-axis.

No.	Operation	Image	Correspondence of coordinates
000	initial state	1 2 3 4	Target[x,y,z]=Source[x,y,z]
001	horizontal flip	2 1 4 3	Target[x,y,z]=Source[sx-x,y,z]
010	vertical flip	3 4 1 2	Target[x,y,z]=Source[x,sy-y,z]
011	Rotate 180° clockwise	4 3 2 1	Target[x,y,z]=Source[sx-x,sy-y,z]
100	Flip along the upper left − lower right corner	1 3 2 4	Target[x,y,z]=Source[y,x,z]
101	Rotate 90° clockwise	3 1 4 2	Target[x,y,z]=Source[sx-y,x,z]
110	Rotate 270° clockwise	2 4 1 3	Target[x,y,z]=Source[y,sy-x,z]
111	Flip along the bottom left − top right corner	4 2 3 1	Target[x,y,z]=Source[sx-y,sy-x,z]

where, i denotes the orientation category. In our orientation classification setting, we set $C = 8$.

Figure 2 shows the overall structure of the multi-tasking network. The backbone of the network is based on the Unet segmentation model. Since the network is originally designed to generate segmentation results, we keep the segmentation branch unchanged and add a new orientation recognition branch which take the segmentation mask as attention map. A weighted binary cross-entropy function is applied to calculate the multi-label segmentation loss between $\hat{Y}_i$ and Y_i. s is set as 4 when dealing with the CMR image segmentation task, which generates background, right ventricle, left ventricle, myocardium. The multi-label segmentation loss is formulated as below:

$$L_{segmentation}(\hat{Y}, Y) = -\sum_{i}^{s}[Y_i log\hat{Y}_i + (1 - Y_i)log(1 - \hat{Y}_i)]w_i.$$

By weighting Orientation recognition loss and segmentation loss, an integral loss function of the proposed loss function can be obtained,

$$L_{integral} - L_{segmentation} + L_{orientation}.$$

The proposed multitask network starts by learning the optimal segmentation network. Thus, *encoder* and *decoder* are optimized according to $L_{segmentation}$. Once this is complete, both *encoder* and *decoder* are fixed and the parameters of $F_{orientation}$ are re-initialized. Now, $F_{orientation}$ is trained according to $L_{orientation}$. Finally, we fine-tune the segmentation module and orientation recognition module simultaneously to obtain an optimized model on both segmentation task and orientation recognition task.

CMR Image Orientation Adjust Tool. To visualize CMR images and perform image orientation recognition, standardization, save operations, we develop a DNN-based CMR image orientation adjust tool, which is embedded with an orientation recognition network. To shorten the response time of the CMR Orientation Adjust Tool, we replace the multi-tasking network with a simplified 3-layer CNN network. We also adopt a different preprocessing method. Suppose given image-label pair (X_t, Y_t), for each pair of X_t. We denote the maximum gray value as G. Three truncation operations are performed on X_t at thresholds $60\%G, 80\%G, G$ to produce X_{1t}, X_{2t}, X_{3t} respectively. The truncation operation maps the pixel whose gray value higher than the threshold to the threshold gray value. Setting different thresholds enforces the characteristics of the image under different gray value window widths to avoid the influence of extreme gray values. The grayscale histogram equalization is also performed on X_{1t}, X_{2t}, X_{3t} to obtain $X'_{1t}, X'_{2t}, X'_{3t}$. We found that the equalization preprocessing of the gray histogram can make the model converge more stably during training. We denote the concatenated 3-channel image $[X'_{1t}, X'_{2t}, X'_{3t}]$ as X'. The orientation recognition CNN only retains the orientation recognition branch while keeping the image-orientation generation steps unchanged. Figure 2 presents the pipeline of the proposed CMR image orientation adjust tool.

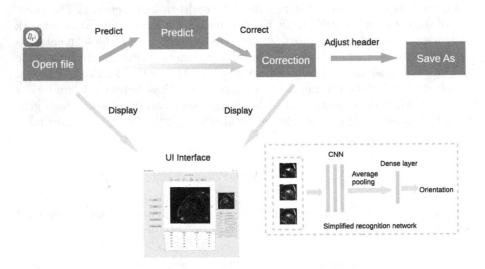

Fig. 2. The pipeline of the proposed CMR image orientation adjust tool.

When adapting the proposed orientation recognition network from a single modality to other modalities, we adopt a transfer learning method to obtain the transferred model. For example, we pre-train model on the balanced-Steady State Free Precession (bSSFP) cine dataset and then transfer model to late gadolinium enhancement (LGE) CMR or the T2-weighted CMR dataset. We first fix the network parameters except for the encoder and retrain the connected layer on the new modality dataset. We go to the next fine-tune step until the model converges. In the fine-tune training, we retrain the encoder and fully connected layer simultaneously on the new modality dataset to obtain an adapted model.

The graphical interface version is suitable for visualization and orientation standardization of a single CMR image. The complementary version of the graphical interface version is the command-line tool version, which supports batch orientation standardization operations of CMR images and provides a simple parameter setting method. By specifying a folder, one line of command is enough to identify the orientation of all MRI files in the folder and correct the files with the wrong orientation.

3   Experiment

Experiment Setup. We evaluate our proposed multi-tasking framework and orientation recognition network on the MyoPS dataset [11,12] and ACDC dataset [1]. The MyoPS dataset was similar to the previous challenge data, i.e., multi-sequence CMR segmentation (MS-CMRseg) challenge, of which both provide the three-sequence CMR (LGE, T2, and bSSFP) and three anatomy masks, including myocardium (Myo), left ventricle (LV), and right ventricle (RV). MyoPS further provides two pathology masks (myocardial infarct and edema) from the 45 patients. The ACDC dataset comprises of single modality short-axis cardiac cine-MRIs of 100 subjects from 5 groups - 20 normal controls and 20 each with 4 different cardiac abnormalities. Annotations are provided for LV, Myo and RV for both end-systole (ES) and end-diastole (ED) phases of each subject. We evaluate multi-task network on ACDC dataset. For the simplified orientation recognition network, we train model for single modality on the MyoPS dataset, then transfer the model to other modalities. For each sequence, we resample each slice of each 3d image and the corresponding labels to an in-plane resolution of 1.367 × 1.367 mm. Image slices are cropped or padded to 212 × 212 for multi-task network and resized to 100 × 100. For the simplified orientation recognition network. We divide all slices into three sub-sets, i.e., the training set, validation set, and test set, at the ratio of 80%, 10%, and 10%.

For Multi-tasking Network. The performance of the orientation recognition module was evaluated by using the accuracy between the predicted orientation and the target orientation. Dice Score was used to measure the accuracy of segmentation. Dice score is an ensemble similarity measurement function, which is usually used to calculate the similarity of two samples. For the predicted segmentation result $\hat{Y}$ and ground truth Y, the dice score is formulated as follows,

$$s = \frac{2|\hat{Y} \cap Y|}{|\hat{Y}| + |Y|}.$$

Table 2. Segmentation Dice Score and orientation recognition accuracy for multitasking network. Note: reported dice score are the average (standard deviation in parenthesis).

Tasks	Segmentation (Dice)			Orientation recognition
	LV	MYO	RV	Accuracy
Results	0.920(0.11)	0.853(0.05)	0.757(0.14)	0.987

Table 2 presents quantitative results of our proposed multi-task network. It can be observed that the proposed method achieves a good segmentation result, with the average dice score of 0.843. The accuracy of orientation recognition reaches 0.986. The quantitative results prove that the proposed multi-task model can effectively deal with the segmentation and orientation recognition tasks simultaneously.

For Simplified Orientation Recognition Network. In each training iteration, a batch of the three-channel images X' is fed into the simplified orientation recognition network (see Fig. 2). Then, the network outputs the predicted orientation network, which is denoted as a 1×3 vector. The predicted orientation is then fed into the standardization module of the CMR image orientation adjust tool. The inverse operation of the predicted orientation error is performed, which warps the MR image to the correct orientation. Then, we perform the same inverse operation on the Orientation vector in the MR file header to obtain the MR file with correct orientation.

Table 3. Orientation recognition accuracy of 2D MS-CMR.

Modality	Accuracy	Description
bSSFP	0.990	Pre-train
LGE	0.852	Transfer learning
T2	0.980	Transfer learning

Table 3 shows the average accuracy on the data set. The description indicates whether the model was trained on this modality or was transferred from other modalities. The high accuracy results provide us with the necessary conditions for the development of the CMR image orientation adjust tool.

4 Conclusion

We have proposed a multi-tasking framework for multi-sequence MRI images that deal with segmentation and orientation recognition tasks simultaneously. Also, we have developed the CMR Orientation Adjust tool (CMRadjustNet), which is embedded with a simplified orientation recognition network. The experiment demonstrates that the embedded orientation recognition network is capable of recognizing the orientation classification from multi-sequence CMR images. Our future research aims to expand the categorization of the CMR image orientation, and study orientation standardization on 3D MRI images.

References

1. Bernard, O., et al.: Deep learning techniques for automatic MRI cardiac multi-structures segmentation and diagnosis: is the problem solved? IEEE Trans. Med. imaging **37**(11), 2514–2525 (2018)
2. Ding, W., Li, L., Zhuang, X., Huang, L.: Cross-modality multi-atlas segmentation using deep neural networks. In: Martel, A.L., et al. (eds.) MICCAI 2020, Part III. LNCS, vol. 12263, pp. 233–242. Springer, Cham (2020). https://doi.org/10.1007/978-3-030-59716-0_23
3. Duan, J., et al.: Automatic 3D bi-ventricular segmentation of cardiac images by a shape-refined multi-task deep learning approach. IEEE Trans. Med. Imaging **38**(9), 2151–2164 (2019)
4. He, T., Guo, J., Wang, J., Xu, X., Yi, Z.: Multi-task learning for the segmentation of thoracic organs at risk in CT images. In: SegTHOR@ISBI (2019)
5. Li, L., Weng, X., Schnabel, J.A., Zhuang, X.: Joint left atrial segmentation and scar quantification based on a DNN with spatial encoding and shape attention. In: International Conference on Medical Image Computing and Computer-Assisted Intervention (2020)
6. Li, L., Zimmer, V.A., Schnabel, J.A., Zhuang, X.: AtrialJSQnet: a new framework for joint segmentation and quantification of left atrium and scars incorporating spatial and shape information. arXiv preprint arXiv:2008.04729 (2020)
7. Vigneault, D.M., Xie, W., Ho, C.Y., Bluemke, D.A., Noble, J.A.: ω-net (omega-net): fully automatic, multi-view cardiac mr detection, orientation, and segmentation with deep neural networks. Med. Image Anal. **48**, 95–106 (2018)
8. Wolterink, J.M., van Hamersvelt, R.W., Viergever, M.A., Leiner, T., Išgum, I.: Coronary artery centerline extraction in cardiac CT angiography using a CNN-based orientation classifier. Med. Image Anal. **51**, 46–60 (2019)
9. Xue, W., Lum, A., Mercado, A., Landis, M., Warrington, J., Li, S.: Full quantification of left ventricle via deep multitask learning network respecting intra- and inter-task relatedness. In: Descoteaux, M., Maier-Hein, L., Franz, A., Jannin, P., Collins, D.L., Duchesne, S. (eds.) MICCAI 2017. LNCS, vol. 10435, pp. 276–284. Springer, Cham (2017). https://doi.org/10.1007/978-3-319-66179-7_32
10. Yue, Q., Luo, X., Ye, Q., Xu, L., Zhuang, X.: Cardiac segmentation from LGE MRI using deep neural network incorporating shape and spatial priors. In: Shen, D., et al. (eds.) MICCAI 2019, Part II. LNCS, vol. 11765, pp. 559–567. Springer, Cham (2019). https://doi.org/10.1007/978-3-030-32245-8_62

11. Zhuang, X.: Multivariate mixture model for cardiac segmentation from multi-sequence MRI. In: Ourselin, S., Joskowicz, L., Sabuncu, M.R., Unal, G., Wells, W. (eds.) MICCAI 2016, Part II. LNCS, vol. 9901, pp. 581–588. Springer, Cham (2016). https://doi.org/10.1007/978-3-319-46723-8_67
12. Zhuang, X., et al.: Multivariate mixture model for myocardial segmentation combining multi-source images. IEEE Trans. Pattern Anal. Mach. Intell. 41(12), 2933–2946 (2019)
13. Zhuang, X., et al.: Evaluation of algorithms for multi-modality whole heart segmentation: an open-access grand challenge. Med. Image Anal. 58, 101537 (2019)

Correction to: Two-Stage Method for Segmentation of the Myocardial Scars and Edema on Multi-sequence Cardiac Magnetic Resonance

Yanfei Liu, Maodan Zhang, Qi Zhan, Dongdong Gu, and Guocai Liu

Correction to:
Chapter "Two-Stage Method for Segmentation
of the Myocardial Scars and Edema on Multi-sequence
Cardiac Magnetic Resonance" in: X. Zhuang and L. Li (Eds.):
Myocardial Pathology Segmentation Combining Multi-Sequence
Cardiac Magnetic Resonance Images, **LNCS 12554,**
https://doi.org/10.1007/978-3-030-65651-5_3

The original version of this chapter was revised. The institute Shanghai United Imaging Intelligence, Co., Ltd., Shanghai, China of the author Dongdong Gu has been removed and the acknowledgement has been changed to "This work was supported by the National Natural Science Foundation of China (61671204)."

The updated version of this chapter can be found at
https://doi.org/10.1007/978-3-030-65651-5_3

© Springer Nature Switzerland AG 2021
X. Zhuang and L. Li (Eds.): MyoPS 2020, LNCS 12554, p. C1, 2021.
https://doi.org/10.1007/978-3-030-65651-5_17

Author Index

Print in the United States
By Bookmasters

Printed in the United States
By Bookmasters